Civil Engineering Heritage London and the Thames Valley

Edited by

Denis Smith, PhD, MSc, DIC, CEng

Other books in the Civil Engineering Heritage Series:
Eastern and Central England. Edited by E. A. Labrum
Ireland. Edited by R. C. Cox and M. H. Gould
Northern England. Edited by R. W. Rennison
Southern England. Edited by R. A. Otter
Wales and West Central England. Edited by R. Cragg

Future titles in the series:
Scotland

London and the Thames Valley
Published for the Institution of Civil Engineers by Thomas Telford Ltd, Thomas Telford House, 1 Heron Quay, London E14 4JD

First published 2001

A CIP record exists for this book
ISBN 07277 2876 8

Typeset by Helius, Brighton and Rochester
Printed in Great Britain by MPG Books, Bodmin

Preface

In 1971 the Institution of Civil Engineers formed the Panel for Historical Engineering Works, its object being to identify, assess and record details of structures deemed to be of historical interest and significance.

To publicise its work, and to stimulate interest in the history both of the profession and the history of engineering and technology, the Institution published the first volume in the *Civil Engineering Heritage* series in 1981.

I would like to acknowledge the assistance given by the Institution's Library staff, notably Mike Chrimes, Head Librarian, and Carol Morgan, the Institution's Archivist, and the courteous and helpful assistance of many others. This book also increased the already large workload of Peter Stephens, Technical Secretary to the Panel for Historical Engineering Works, and I am most grateful for his knowledge of London railway history and for his always cheerful co-operation. I am also indebted to Malcolm Tucker for finding time in his busy schedule to share his comprehensive knowledge and in particular for his input to Chapter 7. I am extremely grateful to Wendie Teppett who kindly and expertly undertook a major photographic survey for the book.

Brian Powell, the Panel member representing the Chilterns and Thames Valley, has undertaken Chapter 9 and illustrated it with his fine photographs. Brian was ably assisted by Robin Sweetnam, David Leiserach and Steve Cummings.

Denis Smith

Contents

Front cover: River Thames (Simmons, Aerofilm Ltd.)

Title page: Three Bridges, Hanwell (Denis Smith)

Metric Equivalents

Imperial measurements have generally been adopted to give the dimensions of the works described, as this system was used in the design of the great majority of them. Where modern structures have been designed to the metric system, these units have been used in the text.

The following are the metric equivalents of the Imperial units used.

Length	1 inch = 25.4 millimetres
	1 foot = 0.3048 metre
	1 yard = 0.9144 metre
	1 mile = 1.609 kilometres
Area	1 square inch = 645.2 square millimetres
	1 square foot = 0.0929 square metre
	1 acre = 0.4047 hectare
	1 square mile= 259 hectares
Volume	1 gallon = 4.546 litres
	1 million gallons = 4546 cubic metres
	1 cubic yard = 0.7646 cubic metre
Mass	1 pound = 0.4536 kilogram
	1 Imperial ton = 1.016 tonnes
Power	1 horsepower (hp) = 0.7457 kilowatt
Pressure	1 pound force per square inch = 0.06895 bar

LONDON AND THE THAMES VALLEY

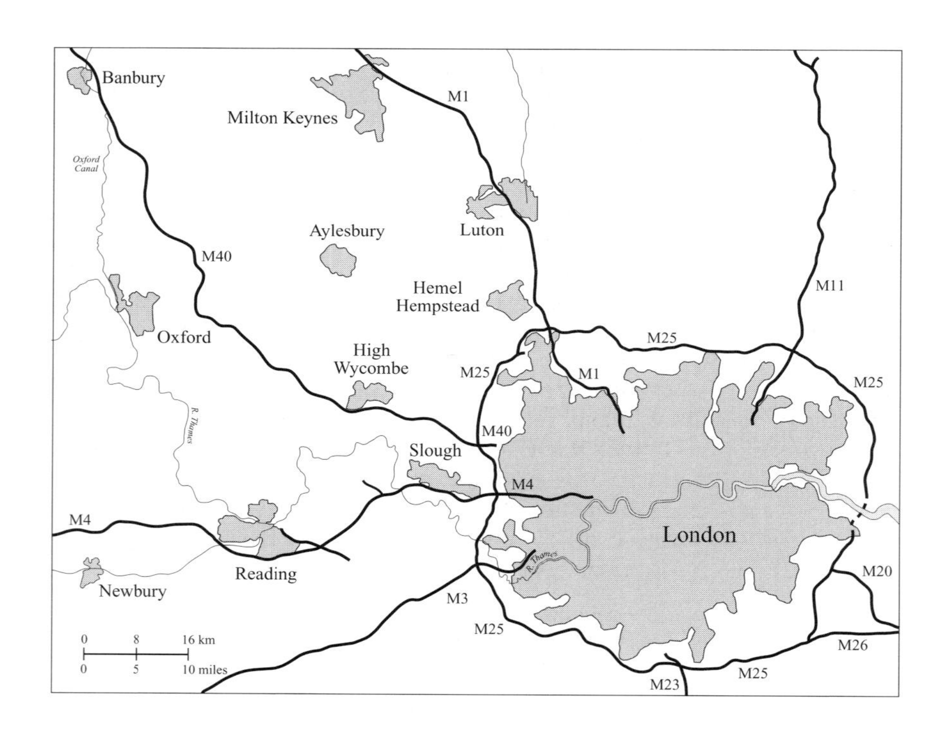

Introduction

The title of this book may conjure up an idea of the vast area covered. Greater London and the Thames Valley have a rich heritage relating to civil engineering and it has not been possible to include everything of merit. The usual selection criteria adopted by the Panel for Historical Engineering Works has included structures of intrinsic engineering interest; a development of a design or construction technique, the first, or an early, use of a new material or combination of materials, and the development of a new structural form. Other, less tangible, issues include the aesthetic value of the work itself and its relationship with its environment, and last, but not least, its association with an eminent engineer, architect or contractor. All these factors have influenced the selection of items in this volume, and this selection has not been easy, involving as it does a blend of these often somewhat subjective value judgements.

Naturally, the Thames itself became a central theme in this volume. The river has generated a prodigious amount of civil engineering work over the centuries and has involved many of the great civil engineers of this country and some from abroad. The poet T. S. Eliot wrote evocatively of an American river describing it as

> ... sullen, untamed and intractable,
> Patient to some degree, at first recognised as a frontier;
> Useful, untrustworthy, as a conveyor of commerce;
> Then only a problem confronting the builder of bridges.
> The problem once solved ... is almost forgotten
> By the dwellers in cities ...[1]

These lines could equally describe the Thames, and it is true that many Londoners who daily use engineering infrastructure take it for granted, at least until something goes wrong. This volume necessarily deals with the design, construction and operation of works that are essential to keeping a large city running 24 hours a day for 52 weeks of the year.

The nature of the engineering heritage of London and the Thames Valley has changed considerably in the past three decades. Steam power has greatly diminished, the hot metal trades are now almost solely represented by the Whitechapel Bell Foundry, docks have closed, some preserved as part of the leisure industry, North Sea gas has closed the London gasworks, and the hydraulic power mains are empty of high-pressure water. An increasingly important part of London's engineering heritage involves the intelligent adaptive re-use of redundant buildings. Nevertheless, much does remain and some of the exciting new transport systems, structures and buildings could become historic works—we have a relatively new flood barrier, a new inner city airport and a new light rail system, amongst others.

Chapter 9 deals with the civil engineering heritage in the Thames Valley, much of which is by contrast in a more rural setting, including parts of Berkshire, Buckinghamshire and Oxfordshire.

[1]Cited in ADAMS, A. (compiler), *Thames: An Anthology of River Poems*. Enitharmon Press, London, 1999, 9.

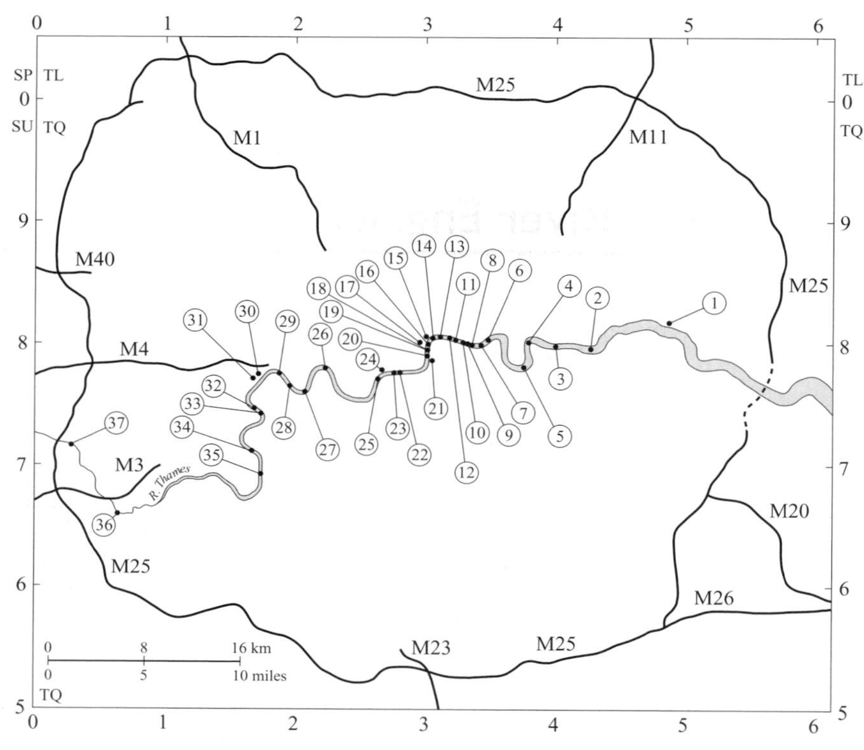

1. Dagenham Breach
2. Woolwich Ferry and Foot Tunnel
3. The Thames Flood Barrier
4. The Blackwall Tunnels
5. Greenwich Foot Tunnel
6. Rotherhithe Tunnel
7. The Thames Tunnel
8. Tower Bridge
9. The Tower Subway
10. London Bridge
11. Cannon Street Bridge
12. Southwark Bridge
13. Blackfriars Rail Bridges
14. Blackfriars Bridge
15. Victoria Embankment
16. Waterloo Bridge
17. Cleopatra's Needle
18. Hungerford Suspension Bridge
19. Charing Cross (Hungerford) Rail Bridge
20. Westminster Bridge
21. The Albert Embankment
22. Grosvenor (Victoria) Bridge
23. Chelsea Suspension Bridge
24. Chelsea Embankment
25. Albert Bridge
26. Hammersmith Bridge
27. Barnes Rail Bridge
28. Chiswick Bridge
29. Kew Rail Bridge
30. Brentford Dock and Augustus Way Bridge
31. Footbridge, Syon Park, Isleworth
32. Richmond Sluice and Half-Tide Lock
33. Twickenham Bridge
34. Teddington Lock and Weir
35. Kingston Bridge
36. Chertsey Bridge
37. Staines Bridge

1. Thames River Engineering

The river Thames has a drainage catchment of about 5264 square miles; it supplies London with water, and has been the great commercial highway for the import and export of goods. To keep the Thames navigable, to prevent it from flooding and to enable it to be easily crossed, has engaged the skills of generations of engineers. This has led to the dredging and marking of navigable channels and the design, construction and management of jetties, ferries, bridges, tunnels, embankments and sluices. This activity has involved the skills of many of the greatest civil engineers of this country and some from abroad. The bodies with responsibility for this range of work have included Trinity House, Thames Commissioners, the Thames Conservancy, the Port of London Authority, the Corporation of the City of London, the Metropolitan Board of Works, the London County Council and the Greater London Council. Their archives are prime sources of information for the engineering historian studying the Thames. In addition, the reports of various Consulting Engineers, Parliamentary Select Committees and of *ad hoc* Commissioners are indispensable.

Rowed ferry boats had existed on the Thames certainly since Roman times, and the ferry from Dowgate in the City to Southwark (on the line of Watling Street) survived until old London Bridge was built. Street names such as *Horseferry Road* in Westminster, and *Horseferry Place* in Greenwich are reminders of ferries with prescribed landing places. A horse ferry is one that could accommodate animals and carts or carriages. These ferries were not without accidents, and the *London Daily Advertiser* of 23 October 1751 reported 'Yesterday a coach and four being taken over in the boat at Twickenham Ferry, the horses took fright and leapt into the water, drawing the coach after them'. *Horseferry Place* in Greenwich is on the alignment of an old crossing to the Isle of Dogs, with a landing onto East Ferry Road. The Horseferry Road crossing from Westminster to Lambeth survived until Westminster Bridge was built in 1750. These ancient ferries were all established by a grant from the Crown and the rights were often transferred in families through several generations. When a new bridge replaced an ancient ferry the enabling

Act of Parliament always included a clause for the 'saving of rights' of the ferrymen. In the late nineteenth century three steam ferries were built. The first was built by private enterprise and plied between Wapping and Rotherhithe. This historic crossing site, on the line of the subsequent Thames Tunnel of Sir Marc Brunel, was first forded by a causeway when the Thames was wide and shallow here. In 1755 a horse ferry was established by an Act of Parliament (28 Geo.2 c.43). The Thames Steam Ferry Company was formed in 1874; it was the first on the Thames, and its route was directly over the Thames Tunnel. The landing jetties were sited adjacent to the London Docks in Wapping and the Surrey Commercial Docks in Rotherhithe. The iron ferry boats were 82 ft long and 42 ft wide, with two paddle-wheels, each independently driven by two 30 hp steam engines. The landing piers comprised a hydraulically operated rise-and-fall staging between two rows of four cast-iron columns. Off the end of the jetty two cast-iron cylindrical dolphins, each 5 ft 6 in. diameter, guided the ferry boats to engage with the staging, at right angles to the shore. The boats were designed and built by Edwards & Symes of Cubitt Town, and the engines by Maudslay Sons & Field. The ferry was opened in October 1877, but operational difficulties were encountered in bringing the boats end-on to either shore, and the ferry failed as a commercial undertaking and closed in 1886. Another private enterprise steam ferry was built connecting Greenwich with the Isle of Dogs. It was on the old horse-ferry alignment, was opened in 1888, but was not a commercial success. The third steam ferry was promoted and designed by the Metropolitan Board of Works and was at Woolwich.

Thames Road Bridges

The building of road bridges spanning the Thames has been undertaken by a variety of promoters over many centuries. These have included the Church, Government Commissioners, The Bridge House Estates of the City of London, and the Metropolitan Board of Works, but principally bridges have been built by companies charging tolls to repay costs and a dividend to shareholders. The work has involved many engineers including Charles Labelye, Robert Mylne, John Rennie, James Walker, I. K. Brunel, Sir John Wolfe Barry, H. M. Brunel, R. M. Ordish, Thomas Page and P. W. Barlow. The increase in traffic by the middle of the nineteenth century led the Government to appoint a Select Committee on Metropolitan Bridges, which reported in 1854. They recommended the abolition of all bridge tolls, the improvement of some bridges, the demolition and replacement of Blackfriars Bridge and the building of four new bridges. Interestingly, they also recommended that Marc Brunel's Thames Tunnel be opened for road traffic—but it was not be.

Table 1: Thames road bridges

Bridge	Engineer	Year
Tower	John Wolfe Barry and Henry Marc Brunel	1894
London	Wm. Holford & Partners	1972
Southwark	Mott, Hay & Anderson	1921
Blackfriars	Joseph Cubitt	1869
Waterloo	Rendel, Palmer & Tritton	1942
Westminster	Thomas Page	1862
Lambeth	Sir George Humphries	1932
Vauxhall	Sir Maurice Fitzmaurice	1906
Chelsea	Rendel, Palmer & Tritton	1937
Albert	Rowland M. Ordish	1873
Battersea	Sir Joseph W. Bazalgette	1890
Wandsworth	Sir Pierson Frank	1940
Putney	J. W. Bazalgette	1886
Hammersmith	J. W. Bazalgette	1887
Chiswick	Alfred Dryland	1933
Kew	Sir John Wolfe Barry & C. A. Brereton	1903
Twickenham	Alfred Dryland	1933
Richmond	Kenton Couse & James Paine	1777
Kingston	Edward Lapidge	1828
Hampton Court	W. P. Robinson	1931
Chertsey	James Paine	1785
Staines	George Rennie	1832

Thames Rail Bridges

Railway companies also promoted bridge structures across the Thames. These bridges were invariably built in metal—cast and wrought iron, or steel. Again eminent engineers were employed including Joseph Locke, John Fowler, Sir John Hawkshaw, Joseph Cubitt, Sir John Wolfe Barry and H. M. Brunel. The bridges in central London were largely built to give access to the city to railways which had arrived in the capital on the south bank.

Over the centuries the management of the regime of the Thames has resulted in a prodigious amount of legislation. The first Act was in the reign of Henry VI in 1423, and from then to 1829 a further 37 Acts were

Table 2: Thames railway bridges

Bridge	Engineer	Date
Cannon Street	Sir J. Hawkshaw	1866
Blackfriars	Sir J. Wolfe Barry and H. M. Brunel	1886
Hungerford	Sir J. Hawkshaw	1864
Grosvenor	Freeman Fox & Partners	1967
Battersea	B. Baker and T. H. Bertram	1863
Putney	W. H. Thomas and W. Jacomb	1889
Barnes	J. Locke and T. Brassey	1849
Kew	W. R. Galbraith	1869
Richmond	J. W. Jacomb-Hood	1908
Kingston	J. W. Jacomb-Hood	*c.* 1900+
Staines	J. Gardner	1856

passed, mostly dealing with tolls. But some deal with the administrative apparatus of managing the Thames, the three most important being the Act of 1771 (II Geo.III c.45) establishing the Thames Commissioners, the Act of 1857 (20 & 21 Vict. c.147) forming the Thames Conservancy, and that of 1908 which transferred the management of the river below Teddington Lock to the new Port of London Authority from 1909.

In 1879 the Metropolis Management (Thames River prevention of floods) Amendment Act was passed making the Metropolitan Board of Works responsible for flood protection works, and these powers were subsequently transferred to the London County Council and then to the Greater London Council and are now vested in the National Rivers Authority.

The sites mentioned in this chapter are described in geographical order from east to west.

THACKER, F. S. *The Thames Highway*: Vol. 1, *General History* (1914); Vol. 2, *Locks and Weirs* (1920).

BROWN, A. Recent types of ferry steamers. *Min. Proc. Instn Civ. Engrs*, 1894, **118**, Pt. 4, 256–70.

REDMAN, J. B. The River Thames. *Min. Proc. Instn Civ. Engrs*, 1876–77, **49**, Pt. III, 67–159.

SKEMPTON, A. W. Engineering on the Thames navigation 1770–1845. *Trans. Newcomen Soc.*, 1983–84, **55**, 153–76.

1. Dagenham Breach

HEW 2193
TQ 495 815

The Thames at this site had breached its bank on several occasions, flooding thousands of acres of the Dagenham and Havering Levels. A breach here was recorded as early as 1376. The land then belonged to Barking Abbey, and in 1621 Cornelius Vermuyden was called in to advise on remedial works at yet another breach. But the large breach, which became a major civil engineering problem, occurred on 17 December 1707. In 1713 Captain John Perry (1670–1732), who had just returned from hydraulic

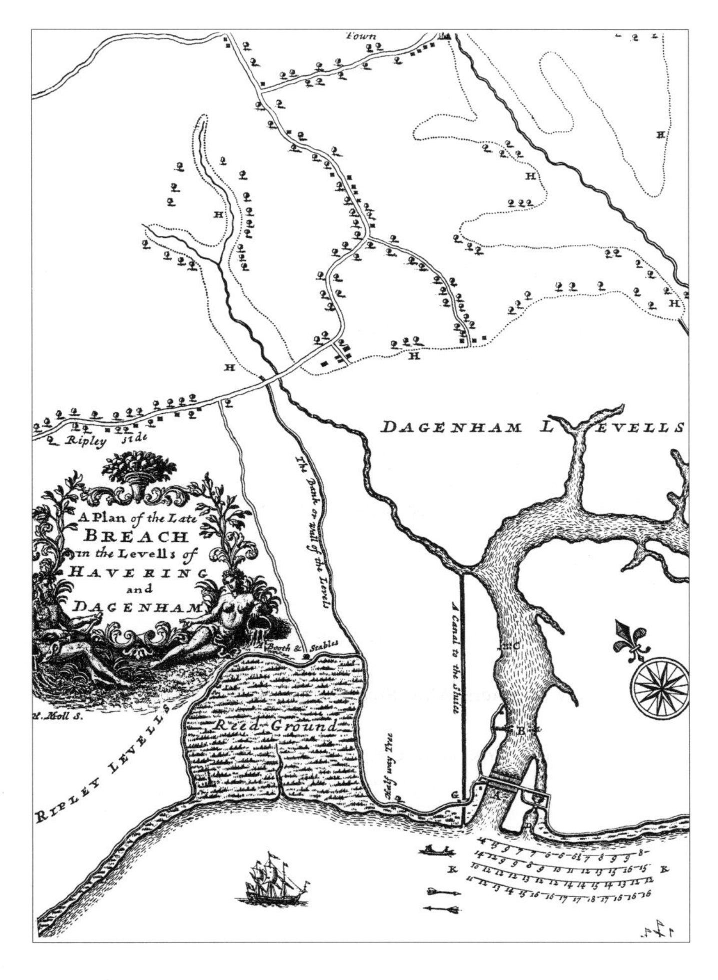

Dagenham Breach

Title page of Captain Perry's account

AN

ACCOUNT

OF THE

STOPPING

OF

DAGGENHAM BREACH:

With the ACCIDENTS that have attended the ſame from the firſt UNDERTAKING.

CONTAINING ALSO

Proper RULES for performing any the like WORK: And PROPOSALS for rendering the Ports of DOVER and DUBLIN (which the Author has been employ'd to Survey) Commodious for Entertaining large SHIPS.

To which is PREFIX'D,

A Plan of the LEVELS which were over-flow'd by the BREACH.

By Capt. JOHN PERRY.

and other works in Russia for Czar Peter, was invited by the landowners to inspect their efforts in repairing the breach. He condemned their work. In 1714 an Act was obtained and Trustees were empowered to supervise the work and they invited tenders. Two were received; Perry quoted £24 000, and William Boswell, who had already made an earlier attempt, estimated £16 500. Boswell's tender was accepted, but his work was unsuccessful and the work was abandoned at the end of 1715. Perry was then asked to undertake the work and he began work on site in the summer of 1716. The works involved Perry, five assistants and about 300 workmen. Although fraught with many setbacks Perry was finally successful. The scheme comprised a substantial earth dam with a row of dovetailed sheet piles at its core with puddled clay on either side. However, his success at Dagenham led to drainage work in Lincolnshire, notably on Deeping Fen. He died in Spalding and is buried in the Parish church. In 1855 an Act was obtained to develop the landlocked lake as a commercial dock promoted by George Remington with Sir John Rennie as engineer, but nothing came of this. After further Acts, work began in May 1865, but ceased in March 1866 when the contractors, Rigby, got into financial difficulties. Henry Ford chose this site for his British car factory and the large lake just north of the factory remains as part of the Dagenham Breach water. In addition, just to the west of the factory is the small Dagenham dock, which is close to Perry's dam site. In 1980 Sir Robert McAlpine & Sons, working on a Tidal Defence Contract at this site for Thames Water, excavated some old dovetailed timber piles. These were carefully removed and taken, for conservation treatment, to the Science Museum where they remain in store.

LYSONS, D. *The Environs of London*. London, 1795.

PERRY, CAPT. J. *An Account of the Stopping of Daggenham Breach*. London, 1721.

HAMILTON, S. B. Captain John Perry (1670–1732). *Trans. Newcomen Soc.*, 1949–51, **27**, 241–53.

2. Woolwich Ferry and Foot Tunnel

HEW 2199
TQ 433 795

There has been a ferry at Woolwich since the thirteenth century. In 1810 the Army established a ferry here, and in 1811 the Woolwich Ferry Company was formed. In August 1885 the Metropolitan Board of Works obtained powers for 'the establishing and regulating of a Ferry across the River Thames at Woolwich'. In December 1886 the Metropolitan Board of Works accepted the Tender of Sir William Armstrong Mitchell and Co. to supply two steam ferry boats at a cost of £10 650 each, and in September 1887 the civil engineering contract was let to Mowlem & Co. for £54 900. The contractors began work on site in December 1887 on the 100 ft by 83 ft landing stages. The Thames Ironworks constructed the bridges and pontoons and Easton & Anderson provided the hydraulic machinery that operated the ramps. The ferry was opened on 23 March 1889 with the paddle steamer *Gordon* in service—her sister vessel, the *Duncan*, was not available until 20 April. Unfortunately, although Sir Joseph Bazalgette was the engineer in charge of the design and construction, the Metropolitan Board of Works was replaced by the London County Council just two days earlier and it was Lord Rosebery, the London County Council Chairman, who performed the opening ceremony. In 1964–66 new reinforced-concrete terminals were built by Marples Ridgeway and Partners to designs by Husband & Co., with trussed steel ramps operating over a 30 ft tidal range.

In 1876 J. H. Greathead began a foot tunnel here but it was not completed. The present Foot Tunnel was designed by Sir Maurice Fitzmaurice for the London County Council, 1909–12. An Act was obtained in 1909, and in March 1910 a contract for the construction of the tunnel was let to Walter Scott and Middleton for £78 860. The tunnel comprises a cast-iron tube of 12 ft 8 in. outside diameter connecting two vertical shafts. Construction of the north shaft began on 1 May 1910 and tunnelling began on 1 December. The length between shaft centres is 1655 ft. The tunnel was excavated by hand labour with the aid of a shield, and a fair day's progress was five rings, or 8 ft 4 in., during 24 hours, the men working three eight-hour shifts.

TABOR, E. H. Woolwich Footway Tunnel. *Proc. Instn Mun. County Engrs*, 1910–11, **37**, 563–66.

3. The Thames Flood Barrier

HEW 2198
TQ 415 796

The low-lying land adjacent to the Thames that had been developed over the centuries as London grew was always prone to flooding. Many serious floods have been recorded in the past, notably that in the year 1236 when Stow records that men rowed boats through Westminster Hall. Flood prevention had been discussed over hundreds of years, and in 1935 the Port of London Authority considered two schemes for dams across the Thames; one at London Bridge costing £800 000 and another at

GREATER LONDON COUNCIL

Thames Flood Barrier

Woolwich, the preferred site, costing £2 million. But it was the disastrous east-coast floods of 1953 that brought the issue to a head, and lent urgency to the problem of finding a method of protecting London from the Thames. Low air pressure combined with gale-force winds were known to produce the storm surge phenomenon where high water levels significantly exceed those predicted. The Greater London Council was the Statutory body responsible for flood protection from the 1960s and the area at risk lying below the level reached in the river in 1953, within the Greater London Council area, was about 45 square miles. The consequences of such a flood were unthinkable. The Greater London Council appointed Rendel, Palmer and Tritton as Consulting Engineers, who reported on a number of schemes in 1958. The most attractive scheme at that time was for a drop-gate structure with two large navigation openings of 500 ft and two subsidiary spans of 250 ft. This scheme was rejected by the Navigation Authorities, although not on engineering grounds. An impasse had been reached in 1965 and the Government appointed Herman Bondi, a mathematician at Kings College, London, to report on the whole question. This he did in 1967, saying that improved flood banks to the river were required and that an upstream barrier would require smaller openings and be more economic. The Greater London Council set up an investigation team in 1968 and a large tidal hydraulic model (1:600 horizontal scale and 1:60 vertical scale) of the Thames from Teddington Weir to Southend was constructed. In addition, hydrodynamic and structural models were built. By the autumn of 1969 the decision was made to site the barrier in Woolwich Reach with downstream improved flood defences from there to the outer estuary. The work was carried out under the Thames Barrier and Flood Prevention Act of 1972. The selected barrier scheme was for a structure with four main navigational openings of 200 ft width each, two small navigational openings of 100 ft and four further openings to allow the tide to flow freely when the structure is open. The gates of the four main navigational spans are of the segmental rising sector type. They sit in recesses in the river bed when open to river traffic, and the barrier is closed by rotating these segments up into the vertical position. The gates are designed to resist a maximum differential head of 30 ft, at which time each gate transfers a thrust of 9000 tons to the piers. This load requires massive structures, and the shafts are steel forgings weighing 48 tons. The gates are designed to close in 15 minutes after the pre-arranged flood warning. The main civil engineering contractors were Costain, Tarmac and Hollandsche Beton Maatschappij, and the steel gates and operating machinery were built by the Davy–Cleveland Barrier Consortium. There is a visitor reception building with an audio-visual display theatre on the south side of the Thames. In addition, the Greater London Council itself supervised some 14½ miles of bank-raising works and, apart from the Thames Flood Barrier itself, drop gate barriers were built on the river Roding at Barking Creek (TQ 455 816) and on the river Darent (TQ 542 780) at its junction with the Thames at Dartford Salt Marshes.

GILBERT, S. and HORNER, R. *The Thames Barrier*. Thomas Telford, London, 1984.

HORNER, R. W. Thames flood protection: Thames Barrier. *J. Inst. General Technician Engrs*, 1974, **85**, No. 2.

4. The Blackwall Tunnels

HEW 2202
TQ 384 807 to TQ 390 797

The first attempt to construct a road tunnel here was made by the Metropolitan Board of Works who obtained an Act in 1887. The design, for three parallel tunnels, was made by Sir Joseph William Bazalgette and the work was to be completed within seven years. This was towards the end of the life of the Metropolitan Board of Works, which was due to be replaced by the new London County Council on 1 April 1889. The Board was about to let the contract when the Government prematurely wound up the Metropolitan Board of Works on 21 March 1889. This was the end of Bazalgette's scheme. In June 1890 the London County Council commissioned Benjamin Baker to inspect and report on the compressed air working at the Hudson River tunnel in New York, and at Sarnia in Canada. Baker reported in October, and by 20 November the London County Council Chief Engineer, Alexander R. Binnie, had produced a new single-tunnel design, under the 1887 Act. The tender of S. Pearson & Son (who were building the Hudson River tunnel), of £871 000, was accepted towards the end of 1891 and work began in 1892. The work began by sinking four shafts in steel caissons, 58 ft external diameter, which were built by the Thames Ironworks on Bow Creek. The circular tunnelling shield, weighing 250 tons, was designed by E. W. Moir, the contractor's Agent, and built by Easton & Anderson of Erith. The shield was driven forward by hydraulic rams, and excavation was by hand. As the shield working was under compressed air at 27 lbf/sq. in. above atmosphere, to prevent a blow-out, a layer of clay 10 ft thick and 150 ft wide was laid on the river bed over the line of the tunnel. Six air compressors totaling 1500 hp were used. The tunnel is 6200 ft long from entrance to entrance. The outside diameter of the cast-iron lining is 27 ft, providing a roadway 16 ft wide with a footway on either side. Some 800 men were employed on the work. The tunnel was lit by three rows of incandescent electric lamps in the roof. It was ceremonially opened by HRH the Prince of Wales on Saturday 22 May 1897. The tunnel was one of the first contracts of the London County Council and the new tunnel one of its last engineering projects.

By the 1930s the old tunnel was becoming inadequate and the London County Council obtained an Act in 1938 for a new tunnel. However, the war intervened and construction work did not begin until 1958 with the northern approach. The new tunnel is about 700 ft to the west of the earlier tunnel and is 3852 ft from portal to portal, with an internal diameter of 28 ft 2 in. The consulting engineers for the bored section of the tunnel itself were Mott, Hay & Anderson, and for the open approaches Mr. H. Iroys Hughes. The architect Terry Farrell designed the two ventilation buildings, and the one on the south side is incorporated in the Millennium Dome structure. The new tunnel was opened in 1967 and carries southbound traffic only—the northbound traffic uses the old tunnel.

The Blackwall Tunnel. *The Engineer*, 1895, **80**, 635–38.

HAY, D. and FITZMAURICE, M. The Blackwall Tunnel. *Min. Proc. Instn Civ. Engrs*, 1896–97, **130**, Pt. 4, 50–98.

The Blackwall Tunnel. *The Engineer*, 1897, **83**, 504–08.

KELL, J. and RIDLEY, G. *ICE Proc.*, 1966, **35**, 253–74; 1967, **37**, 537–55.

5. Greenwich Foot Tunnel

HEW 2203
TQ 383 788 to
TQ 783 779

Built from Island Gardens on the Isle of Dogs to the Greenwich waterfront, the tunnel was built to replace a ferry. Two vertical shafts, each of 43 ft external diameter, give access to the tunnel by spiral staircase or lift. The tunnel is 1217 yd long between shaft centres and is made of cast-iron rings of 12 ft 9 in. external diameter. It was built for the London County Council under their Engineer Sir Alexander Binnie, the resident engineer was W. C. Copperthwaite and the contractors were J. Cochrane and Sons. Opened in 1902, it is still in use.

COPPERTHWAITE, W. C. The Greenwich foot tunnel. *Min. Proc. Instn Civ. Engrs*, 1902, **150**, Pt. 4, 1–24.

6. Rotherhithe Tunnel

HEW 2204
TQ 356 857 to
TQ 354 798

The London County Council obtained powers to build this tunnel under the Thames Tunnel (Rotherhithe and Ratcliff) Act of 1900. Construction began in 1904 when four construction shafts were sunk, two on each side of the river. The shafts remain as ventilation shafts, and those on the river's edge, which are of red brick with stone dressings, also provide pedestrian access via spiral iron staircases. The distance between shafts 1 and 4 is 3689 ft and is cast-iron lined. On both sides of the river the road is in open cutting with brick retaining walls followed by cut and cover sections. Entrance to the tunnel on the Surrey side is framed by the cast steel segments of the cutting edge of the tunnel shield, forming in effect a loading gauge for the tunnel. The tunnel was opened in 1908. It was designed by Sir Maurice Fitzmaurice, Engineer to the London County Council, the resident engineer was Edward H. Tabor and the contractors were Price and Reeves.

TABOR, E. H. The Rotherhithe Tunnel. *Min. Proc. Instn Civ. Engrs*, 1908, **175**, 190–251.

7. The Thames Tunnel

HEW 177
TQ 358 798 to
TQ 351 802

An early attempt to tunnel under the river from Rotherhithe to Limehouse was made by Robert Vazie and Richard Trevithick between 1805 and 1809. The project was described as the Thames Archway, and a pilot driftway was attempted from a shaft in Rotherhithe. Simple Cornish mining techniques were used, without a shield, and the project was abandoned when about 1000 ft out of a total of 1200 ft was completed. It was, however, a heroic attempt.

In January 1818 Sir Marc Brunel obtained a patent (No. 4204) for a circular tunnelling shield with a rotary cutter. This was ahead of the technology of the day and was a landmark event in the history of tunnelling. However, the need for a road crossing of the Thames to connect the docks in Wapping and Rotherhithe was obvious and was under discussion at this time. In 1823 Sir Marc Brunel produced a plan for a tunnel from Rotherhithe to Wapping, and in 1824 a group of promoters obtained an

Thames Tunnel, Rotherhithe shaft and pumping station

DENIS SMITH

Act forming the Thames Tunnel Company. Jolliffe and Banks had undertaken a borehole soil investigation in the line of the proposed tunnel. Work began in February 1825 with the setting out of the Rotherhithe shaft. The 50 ft diameter, 40 ft high brick shaft was built above ground and then sunk into its final position. The rectangular tunnelling shield, built by Maudslay in Lambeth, was then assembled at the bottom of the shaft facing the river and the heading towards Wapping began in November 1825. The two archways in the tunnel are contained within a rectangular mass of brickwork 37 ft 6 in. wide and 22 ft 3 in. deep. In May 1827 the first irruption of the river into the workings occurred when the tunnel was 549 ft long. The second irruption, in January 1828, was more severe—six men died, the young I. K. Brunel was severely injured, the shield was damaged and work ceased. There were financial problems, and in August the tunnel was walled up. In 1833 a loan from the Treasury enabled work to re-start. The old shield had to be removed and a new version, built at Rennie's Albion Ironworks, was installed and started moving in February 1836. After two more irruptions of the river Brunel took possession of the Wapping shaft site in June 1840. On 25 March 1843 the 1200 ft tunnel was opened to pedestrians, and spiral staircases in the shafts provided access. The tunnel was never used for vehicular traffic. In May 1865 the East London Railway Company was formed, and in September of that year the tunnel was sold to the railway company for £200 000. A track was laid in each archway and the first train ran through the tunnel on 7 December 1869. The line was electrified, and on 31 March 1913 the Metropolitan Railway began a service over the East London Line. The line became part of London Underground on 29 January 1914, and remains part of the London Underground system today. On 24 March 1995 the whole tunnel was Listed Grade II*. The Thames tunnel is undeniably the most important example of civil engineering heritage in London.

In 1842 a new engine house was built alongside the Rotherhithe shaft to house machinery for draining the tunnel. The building survives as a scheduled Ancient Monument and the *Brunel Exhibition Project* completed restoration work in 1979 and opened the Museum there in June 1980.

SKEMPTON, A. W. and CHRIMES, M. M. Thames Tunnel: geology, site investigation and geotechnical problems. *Géotechnique*, 1994, **44**, 191–216.

CLEMENTS, P. *Marc Isambard Brunel*. Longman, London, 1970.

BEAMISH, R. *Memoir of the Life of Sir Marc Brunel*. Longman, London, 1862.

8. Tower Bridge

HEW 31
TQ 336 802

This is the most famous bridge over the Thames and has become a symbol of London itself. The need for river crossings below London Bridge arose as a result of the development of the dock system downstream, and the 1877 Act, which freed the inner London Thames bridges from

WENDIE TEPPETT

Tower Bridge

Tolls. East Londoners argued that their rates had been used for this purpose and demanded below-bridge crossings. In 1879, 39% of London's population lived east of London bridge. All this led to the building of Woolwich Free Ferry, Blackwall Tunnel and Tower Bridge. A decision was taken that a bridge should be built on a site close to the Tower of London and the rivalry between the Metropolitan Board of Works and the City of London was again apparent. In 1878 Joseph Bazalgette of the Metropolitan Board of Works submitted three designs—two for high-level girder bridges and a magnificent single-span steel arch. But the City of London refused all designs originating outside their own Architect's Department and won the day by saying that the City would defray all the costs from their Bridge House Estates funds. In 1878, Horace Jones, the City Architect, had sketched an outline scheme for a low-level opening bascule bridge. In 1884 John Wolfe Barry was appointed consulting engineer, and in October Barry wrote to Jones agreeing on a bascule bridge with high-level walkway and hydraulic lifts. For this job John Wolfe Barry went into partnership with Henry Marc Brunel. An Act was obtained in August 1885 and work on site began in April 1886 on the first caisson, and in June a memorial stone was laid by HRH the Prince of Wales and Horace Jones was Knighted. In March 1886 John Jackson was appointed contractor for the piers and abutments and, in 1887, for the northern approach works. In July 1888 William Webster was appointed to build the southern approach. The two massive piers in the river are each 70 ft wide and 185 ft 4 in. long providing a clear navigational width of 200 ft. The contractors were required by the Thames Conservancers to maintain a minimum navigational clear width of 160 ft throughout the construction period. This constraint delayed the work in the river considerably. Design of the hydraulic power system for lifting the bascules, detailed design calculations (by H. M. Brunel) for the steel towers and the architectural design of the masonry cladding continued as the river work progressed. As a result it was not until December 1887 that the contract for the hydraulic machinery was let to Sir William Armstrong, Mitchell & Co., and in May 1889 that the contracts for steel superstructure and the masonry cladding were let to Sir William Arrol & Co. of Dalmarnock Ironworks, Glasgow, and to Perry & Co., respectively. By January 1889 both river piers were complete.

Armstrong's hydraulic equipment comprised a pumping station on the Surrey shore in which two double tandem compound steam engines, each of 360 hp, drove force pumps pressurising the water to 700 lbf/sq. in. There were four Lancashire boilers with a working steam pressure of 85 lbf/sq. in. In addition, two hydraulic accumulators with 20 in. diameter rams and 35 ft stroke met the peak demand for pressurised water when opening the bascules. The accumulator tower and the brick chimney are conspicuous landmarks when approaching the bridge from the south. The high pressure water was taken out to the two piers, in which were engine rooms each with two large and two small hydraulic motors driving pinions engaging with a rack quadrant and the end of each bascule.

Arrol's contract for the steel superstructure involved the fabrication and erection of 11 000 tons of steel, 1200 tons of cast iron and 580 tons of lead. The firm's site engineer was J. E. Tuit. The metalwork was to be shipped from Glasgow to London and delivered to contractors' barges half a mile below the bridge site at the rate of 50 to

100 tons per week. No single piece of steelwork weighed more than 5 tons. The four steel columns in each tower were 119 ft 6 in. high, octagonal in plan, fabricated from plates, angles and tees, and were 5 ft 9¾ in. across the flats. The octagonal steel base-plate was 14 ft across and was bedded on a stone base 16 ft square and 3 ft thick. The delivery of all materials was made on the shore side of each river pier.

The bridge, which had cost £1 184 000, and eight lives, was opened with great ceremony by HRH the Prince of Wales on Saturday 30 June 1894. The design was criticised on aesthetic grounds because of the non-load-bearing masonry cladding. During the first month the bridge was opened 655 times and by 1955 a total of 325 358 times. The bridge was converted to electro-hydraulic operation in 1974 and the steam plant was de-commissioned. The consulting engineers were Mott, Hay and Anderson. The towers and the high walkway now contain a multi-media exhibition attracting many visitors.

WELCH, C. *History of the Tower Bridge*. Smith, Elder & Co., London, 1894.

TUIT, J. E. The Tower Bridge. *The Engineer*, 15 December 1893.

9. The Tower Subway

HEW 233
TQ 334 806 to
TQ 333 802

This river crossing runs from Tower Hill in the City to Vine Street on the Surrey shore. It was built under an Act of 1868 (31 & 32 Vict. c.8) which established The Tower Subway Company with three Directors and a capital of £12 000. The engineers were P. W. Barlow and J. H. Greathead with P. W. Barlow Jr. as Resident Engineer. This foot passenger system comprises a 7 ft diameter, 1320 ft long tunnel between the two vertical access shafts. This was the first cast-iron lined tunnel in Britain and it was lined with rings in 18 in. lengths. Its greatest depth is 48 ft below high water. Thomas Tilley, the shaft contractor, began sinking the Tower Hill shaft in February 1869, and the subway opened early in 1870. Passenger access was through the 10 ft 2 in. diameter shafts, which are 60 ft deep. A 4 hp steam engine operated a lift in the shaft together with the cable haulage system for the iron 12-seater omnibus. The journey time was 2½ to 3 minutes and it was 'extensively used by the working classes who were formerly entirely dependent on the ferries'. When Tower Bridge opened in 1894, as a free pedestrian crossing, there was no incentive to pay the toll in the subway and it fell into disuse. In 1897 an Act (60 & 61 Vict. c.97) enabled the Subway to be sold to the London Hydraulic Power Company for £3000. The Subway is still in use as a cable duct tunnel serving the communications industry. The entrance to the tunnel on Tower Hill survives.

The Engineer, 26 March 1869, 225.

Engineering, 12 November 1869, 319

Obituary, P. W. Barlow. *Min. Proc. Instn Civ. Engrs*, 1884–85.

10. London Bridge

HEW 261
TQ 328 806

London Bridge has a history spanning nearly 2000 years. The first bridge over the River Thames was the Roman timber structure built close to the site of the present bridge. Remains of this bridge were excavated in 1981. But the title 'Old London Bridge' (HEW 255) is used to describe the early thirteenth century masonry bridge attributed to Peter, Chaplain of Colechurch. Most documents attribute the beginning of the work to the

London Bridge

WENDIE TEPPETT

year 1176. The new bridge had 19 arches resting on 20 piers; it was built slightly upstream of its timber predecessor, and was completed in 1209. The bridge with its chapel, shops and houses was much altered over the centuries, principally by removing small arches and replacing them with a larger arch, to improve the navigation on the Thames. The bridge was being maintained by the City's Bridge House Estates at least as early as 1283. It is interesting to note that this bridge was the only one in the capital for over 500 years, until Westminster Bridge was built in the mid-eighteenth century.

By the first decade of the nineteenth century old London Bridge was extremely crowded. On one day in June 1811 a traffic census revealed that 89 640 pedestrians, 769 wagons, 2924 carts, 1240 coaches, 485 gigs and 764 horsemen crossed the bridge. At the same time the old bridge was an obstruction to river traffic. In 1820 a Select Committee was established to inquire into the state of the London Bridge; they reported in May 1821 proposing a new bridge. On 4 October of that year John Rennie (1761–1821), who had been involved in the new design, died. The Corporation obtained an Act in 1823 empowering them to demolish old London Bridge and to build a new bridge to John Rennie's design (HEW 260). It was not until 1824 that the younger John Rennie (1794–1874) was appointed Engineer to the scheme. On 4 March a contract was let to Jolliffe and Banks who began work on 15 March. Cofferdams, 165 ft by 115 ft, were constructed in the river. The walls comprised three rows of 14-in. square timber piles driven 25 ft into the river bed. The bridge comprised five semi-elliptical granite arches, the central one being of 152 ft 6 in. span, the second and fourth arches of 140 ft span, and the first and fifth each of 130 ft span. The carriageway was 36 ft wide. Rennie's bridge was built parallel to Old London Bridge and 175 ft upstream of it. The bridge was opened by King William IV on 1 August 1831. One of Rennie's land arches survives on the south bank close to Southwark Cathedral and crossing the alignment of Tooley Street and Montague Close.

Rennie's London Bridge of 1831 was becoming inadequate for modern traffic by the late 1950s and early 1960s, and the Corporation proposed a new bridge (HEW 261) on the same centre line as the older bridge. The London Bridge Act was passed in 1967. Mott, Hay and Anderson were appointed engineers, with William Holford and Partners as architectural advisers, and the contract was let to John Mowlem and Company in 1967 for £4 066 000. Two navigation channels, each 100 ft wide, had to be maintained during all stages of construction. Two slender piers are founded in the river bed. They are constructed of mass concrete faced with granite blocks laid with recessed joints. The spanning elements are four longitudinal box beams constructed of precast concrete segments. The units were precast at the Russia Dock in the Surrey Docks and transported to the bridge site by barge. When in position prestressing strands were threaded through the box units and tensioned at the anchorages. Work started on site in November 1967 and the new London bridge was opened by the Queen in 1972.

In April 1968 the stonework of Rennie's bridge was sold to McCulloch Properties Inc. of California for £1 025 000. The numbered blocks of masonry were shipped to America and have been re-erected at Lake Havasu City, Arizona. The removal of the earlier multi-arched London Bridge was, in effect, the removal of a tidal barrage and this led to increased scour on the foundations of the upstream bridges, necessitating a great deal of remedial work and in some cases the replacement of the affected bridge.

CEMENT AND CONCRETE ASSOCIATION. *London Bridge*. Wexham Spring, 1971.

RENNIE, SIR J. *Autobiography*. Spon, London, 1875.

11. Cannon Street Bridge

HEW 2251
TQ 325 806

The South Eastern Railway Company wished to have an extension of their line from London Bridge into the City, and powers for this work were obtained in 1861. The bridge was designed by Sir John Hawkshaw, Chief Engineer to the South Eastern Railway. The piers are cast-iron cylinders firmly founded on the London Clay. The cylinders were filled with concrete up to the level of the river bed and lined with brickwork above. There are four cylinders to each pier. The cylinders are treated architecturally as fluted Doric columns, the outer ones having Doric capitals. The bridge is formed of heavy longitudinal wrought-iron plate girders each 8 ft 6 in. deep. The two outer girders have double web plates. The bridge was built in 1863–66 and cost £193 000. It was widened on the westerly side by adding two further cylinders to each pier in 1886–93. In 1910–13 the bridge was strengthened to carry heavier class locomotives. In the 1980s it underwent major strengthening works, which included pressure grouting to fill voids between the inner surface of the iron cylinders and the brick lining, the addition of reinforced-concrete collars at the top of the cylinders, and the replacement of the original wrought-iron crossheads by reinforced-concrete beams. The work was completed in 1983.

12. Southwark Bridge

HEW 527
TQ 324 806

In 1813 an Act of Parliament established the Southwark Bridge Company, with John Rennie as engineer, who charged a design and supervision fee of £7500. Rennie's design crossed the Thames in three impressively large cast-iron arches; the centre span was 240 ft and the side spans both 210 ft. Eight parallel cast-iron ribs each comprising 13 voussoirs supported the road and footways. The bridge cost £800 000 and was declared open at midnight on 24 March 1819. The Walker Ironworks of Rotherham supplied the 5780 tons of ironwork at £18 per ton, including casting, transport and erection on site. Jolliffe & Banks were the masonry contractors and the river piers comprised Bramley Fall and Whitby stone. In 1866 the bridge was purchased from the Company, for £200 000, by the Corporation of the City of London and freed from tolls. In June 1909 the journal *Civil Engineering* wrote 'Whether or not Southwark Bridge should be rebuilt has for years past been a hardy subject for gossip in the City'. In 1913 demolition of the Rennie bridge was begun, but was delayed by the 1914–18 war and the present bridge, designed by Basil Mott, of Mott, Hay & Anderson, was eventually opened by King George V on 6 June 1921. The present bridge comprises five spans of steel-plate girder ribs.

13. Blackfriars Rail Bridges

HEW 2252
TQ 317 806

The extended London, Chatham & Dover Railway reached the south bank of the Thames in 1864. Although an Act for building a railway bridge was obtained in 1860 the work was delayed as the City Corporation could not decide on a design for the bridge. However, Joseph Cubitt and F. T. Turner completed the work on 21 December 1864. The bridge had four rail tracks carried on four river spans, two shore spans of 160 ft, two inner spans of 175 ft and a centre span of 185 ft. The wrought-iron girders were of lattice form and 15 ft 6 in. deep. The transverse river piers comprise three cast-iron cylinders, 18 ft in diameter and clad in masonry to high-water level. The cylinders were filled with concrete and capped with granite supporting 5 ft diameter cast-iron columns 21 ft tall. The girders of this bridge were demolished in the 1980s, but the cast-iron Doric columns remain in the river. The Arms of the London, Chatham & Dover Railway, in cast iron, have been restored and are easily visible on the south pier. Stones from Old Westminster Bridge, demolished in 1861, were used in the abutments. When opened the bridge had cost £220 000.

The second railway bridge, immediately downstream of the Cubitt bridge, was known as St. Paul's Bridge and was opened in May 1886. The engineers were Sir J. Wolfe Barry and H. M. Brunel in conjunction with William Mills, the London, Chatham & Dover Railway engineer. The 81 ft wide structure carried seven rail tracks. The bridge has five spans of segmental wrought-iron arches and, for navigational reasons, the three middle river spans match those of the Cubitt bridge alongside, the Surrey shore span being 183 ft, the centre and Middlesex shore spans 185 ft and the second and fourth spans 175 ft. The spandrels have vertical metal posts and horizontal members. The ironwork contractor was the Thames Ironworks and Shipbuilding Company of Blackwall. In 1937 St. Paul's Bridge was merged with Cubitt's rail bridge.

CRUTWELL, G. E. W. The new bridge of the London, Chatham and Dover Railway Company over the Thames at Blackfriars. *Min. Proc. Instn Civ. Engrs*, 1899–90, **101**, Pt. 3, 25–37.

14. Blackfriars Bridge

HEW 2200
TQ 316 806

There have been two road bridges on this site separated by exactly a century, the first in 1769 and the second in 1869. The Corporation of the City of London selected

Robert Mylne, architect and engineer, winner of a competition attracting 69 entries, to design this structure, which was originally called 'Pitt Bridge'. Mylne designed his nine-span masonry arched structure with semi- elliptical arches. His use of this arch profile led to a great deal of public criticism, as previous arch bridges on the Thames had been of segmental form. The central span was 100 ft, the other spans diminishing to 98, 93, 83 and 70 ft. However, his structure was a landmark example of civil engineering design with excellent foundations and with architectural treatment including double pilasters above the cutwaters—a device subsequently adopted by Rennie on the adjacent Waterloo Bridge.

In 1832 the City Corporation commissioned Walker & Burges to survey and report of the condition of Mylne's bridge. Their report confirmed that the structure and foundations were in poor condition, and after spending £105 000 on remedial works the City decided on a new bridge.

The present bridge was designed by Joseph Cubitt with wrought-iron arch ribs. Work began in June 1864. To prevent scour Cubitt sank iron caissons deep into the clay, half filled them with concrete and built the piers in granite-faced brickwork. The central river spans matched those of his adjacent rail bridge. The City wished to ornament this functional structure and insisted on cutwaters of polished red granite surmounted by Portland stone carvings of birds and plants by the sculptor J. B. Philip.

The foundation stone of the bridge was laid in 1865 and the bridge was opened by Queen Victoria on 6 October 1869—she opened Holborn Viaduct on the same day.

15. Victoria Embankment

HEW 2196
TQ 303 796 to
TQ 316 803

Sir Joseph Bazalgette, as Engineer to the Metropolitan Board of Works, fulfilled a long-cherished scheme when he embanked the Thames in central London. The Victoria Embankment runs as a quadrant from Westminster to Blackfriars Bridges and was laid out on James Walker's line proposed in 1841. This embankment is technically the most interesting of the three as, apart from the road above, the structure was designed to

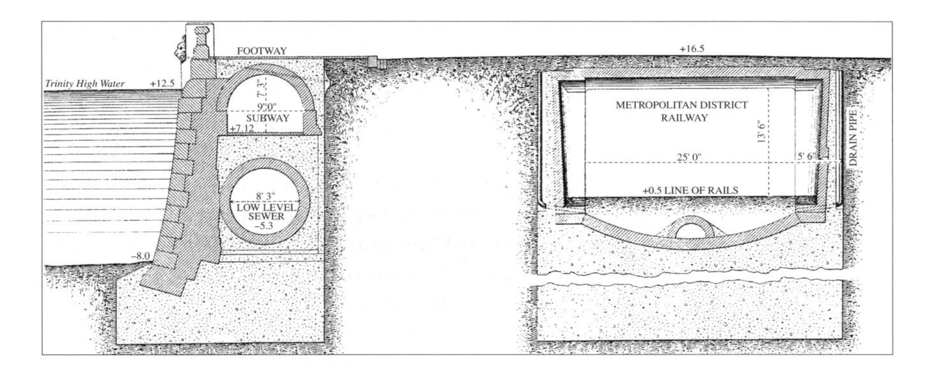

accommodate a gas and water service duct, the northern low-level sewer, and the Metropolitan District Railway. An Act was obtained in 1862 and the work was let under three contracts. No. 1 was the section from Blackfriars to Waterloo Bridge, let to George Furness, No. 2 from Waterloo Bridge to the Temple let to Mr. Ritson, and the remainder (including the whole roadway) to Blackfriars was let to William Webster. Work began in 1864, on No. 1 Contract, by constructing a large wrought-iron caisson cofferdam wall formed of iron cylinders driven into the river bed. Much of the excavation was undertaken by 8 hp steam dredgers carried on a 55 ft wide travelling bridge running on rails mounted on a timber superstructure. The embankment has a curved battered surface and is faced with granite. The 100 ft wide roadway is about 1¼ miles long and the embankment reclaimed just over 37 acres of land. The embankment was opened by HRH the Prince of Wales on 13 July 1870.

Victoria Embankment and Metropolitan District Railway

PORTER, D. H. *The Thames Embankment*. University of Akron Press, Ohio, 1998.

BAZALGETTE, E. The Victoria, Albert, and Chelsea embankments of the River Thames. *Min. Proc. Instn Civ. Engrs*, 1878, **54**, Pt. 4, 1–60.

16. Waterloo Bridge

HEW 262
TQ 308 805

In 1809 an Act was passed incorporating the 'Strand Bridge Company' with capital of £500 000. John Rennie, the elder, was asked to prepare designs for a stone bridge from the Strand to Lambeth, which was renamed Waterloo Bridge (HEW 263) after 1816. Rennie designed a bridge with nine elliptical arches, each of 120 ft span and

35 ft rise. The carriageway was 28 ft wide with two raised footpaths. The cutwaters were 85 ft point to point, and above the cutwaters the piers were surmounted by two Doric columns supporting entablature. The foundation stone was laid on 11 October 1811. The total cost of the bridge was £1 million and it was opened by the Prince Regent on 18 June 1817. Foundation protection work was necessary during 1882–84. In 1923 three piers were found to be sinking, and in June 1934 the bridge was closed and demolition was begun. However, the Second World War delayed the construction of the present bridge.

The need for a new Waterloo Bridge (HEW 262) was clearly urgent if the London County Council was to provide better traffic facilities and at the same time to improve river navigation. But it was not until 1936 that Parliament approved the London County Council's Money Bill. In 1937 tenders were obtained for the new structure and work began in October. But only in 1938 did the Minister of Transport agree to make a Road Fund grant towards its cost.

The bridge is of reinforced concrete, faced with Portland stone, with five spans, each nearly 240 ft clear. The roadway is 58 ft wide with two footpaths, each 11 ft wide. On 11 August 1942 two lanes were opened to road traffic and on 21 December the footpaths were opened and the temporary bridge closed. The Engineers were Rendel, Palmer & Tritton, appointed by the London County Council, in association with Sir Pierson Frank the Council's Chief Engineer. The architect was Sir Giles Gilbert Scott and the contractors Peter Lind & Co. Ltd.

BUCKTON, E. J. and CUEREL, W. The new Waterloo Bridge. *J. Instn Civ. Engrs*, 1942–43, **20**, 145–201.

17. Cleopatra's Needle

HEW 2197
TQ 306 805

This stone obelisk, quarried at Syene, was erected at Heliopolis about 1500 BC. The transportation of the obelisk from Egypt and its erection in London was initiated by members of the Institution of Civil Engineers in 1877. By then there had long been a tradition of moving Egyptian obelisks to Europe, going back to Roman times. The promoters of this scheme were Erasmus Wilson and the brothers John and Waymann Dixon. In 1877 Wilson

ILN

Cleopatra's Needle, lifting rig

offered to pay £10 000 once the obelisk was erected in London, with John Dixon, civil engineer, taking all the risk on a no fee if unsuccessful basis. At this point John Dixon involved Benjamin Baker to design and supervise the engineering aspects of the project. The 68 ft 9 in. obelisk was lying; half-buried in Egypt and was estimated to weigh 186 tons. Baker designed a special vessel to transport the obelisk both on land and sea, this was a wrought-iron cylinder, 15 ft in diameter and 92 ft long, and pointed at both ends. In mid-March 1877 a contract to build the vessel, named the *Cleopatra*, was placed with The Thames Ironworks Company and in August it was ready for use on site in Egypt. The obelisk was wedged securely inside the *Cleopatra* and reached the coast on 8 August, and on 21 September the obelisk began its journey to England being towed by the *Olga*. Baker blithely recorded 'All went well for the first 2400 miles', then the *Cleopatra* became adrift in the Bay of Biscay. The vessel and obelisk were considered lost, but she was later found and restarted her journey to England on 16 January 1878, arriving at Gravesend four days later.

At the chosen site on the Victoria Embankment a 50 ft high timber structure was erected comprising four upright posts, diagonal bracing and raking struts. The obelisk was landed in the horizontal position and the central third was encased in a wrought-iron casing with knife-edged trunnions. The trunnions rested on iron box girders at the ends of which were hydraulic jacks. The Needle was lifted in the horizontal position and was swung effortlessly into the vertical position on 12 September 1878.

ALEXANDER, LT. GENERAL SIR J. *Cleopatra's Needle, the Obelisk of Alexandria*. Chatto & Windus, 1879.

BAKER, B. Cleopatra's Needle. *Min. Proc. Instn. Civ Engrs*, 1879–80, **61**, Pt. 3, 233–43.

Thames Ironworks Gazette, 30 June 1900, 111–15.

LEWIS, M. J. T. Roman methods of transporting and erecting obelisks. *Trans. Newcomen Soc.*, 1984–85, **56**, 87–110.

18. Hungerford Suspension Bridge

HEW 2201
TQ 306 803

This pedestrian bridge was built under an Act of 1836 and an amending Act of 1843, with the object of bringing more custom to the recently rebuilt Hungerford Market on the north bank. The bridge was designed by I. K. Brunel and was opened on 1 May 1845. The bridge comprised masonry towers, with access for riverboat passengers, and two, double, wrought-iron chains. The chain links were 24 ft long and 7 in. by 1 in. cross-section. The overall length was 1342 ft with a central span of 676 ft and the deck was 14 ft wide. The bridge had a difficult access on the south side and was the subject of aesthetic criticism, one publication saying 'Hungerford Bridge places the west end of London in direct communication with the worst part of Lambeth' and 'The piers are of ornamental brickwork, of very questionable taste, and apparently of doubtful solidity'.[1] The bridge was not a financial success and it was under demolition in January 1863 after only 17 years, having cost a little over £100 000. The red brick tower bases survive and were used to support the Charing Cross railway bridge. The chains were salvaged and were taken to Bristol to complete the Clifton Bridge as a memorial to Brunel.

[1]WEALE, J. *London Exhibited*, 1851, 280.

19. Charing Cross (Hungerford) Rail Bridge

HEW 197
TQ 305 807

Built to give the South Eastern Railway access to the north of the river at Charing Cross, this wrought-iron lattice girder bridge was opened in 1864. The designer was Sir John Hawkshaw and the contractor was Cochrane & Co. of Dudley in the West Midlands. The bridge is on the same alignment as Brunel's Hungerford suspension bridge and makes use of the two red brick bases of Brunel's suspension towers. The other piers are cast-iron cylinders which were sunk by excavating inside and loading, taking them down to 70 ft below Trinity High Water (THW). The cylinders are filled with concrete up to the river bed level and lined with brickwork above. The bridge comprises nine spans, three of 100 ft and six of 154 ft. The trusses are at 49 ft 9 in. centres and there is a footbridge on the downstream side for which a replacement is under construction as this book goes to press. A new footbridge on the upstream side is also being built. The rails are 31 ft above THW. There were four tracks originally; now there are six.

HAYTER, H. The Charing Cross Bridge. *Min. Proc. Instn Civ. Engrs*, 1862–63, **22**, 512–39.

20. Westminster Bridge

HEW 264
TQ 305 797

The first bridge on this site was opened in 1750 and was only the second Thames bridge in London for 540 years since Old London Bridge. A suggestion for a bridge here was made in 1664, but was vigorously resisted by Londoners. However, in 1736 an Act was obtained, Commissioners were appointed, and £5 lottery tickets were sold to raise the necessary £625 000. The designer was Charles Labelye, a naturalised Swiss engineer and architect whose appointment in May 1738 raised some resentment amongst English architects. The bridge, 300 ft longer than London Bridge, was built of Portland stone and had 13 large and two smaller arches, each semicircular. The bridge suffered badly from the increased scour of the foundations after the removal of Old London Bridge and several Select Committees and many civil engineers were involved with its maintenance.

WENDIE TEPPETT

Westminster Bridge

By the middle of the nineteenth century it was obvious that a new bridge was required, partly, one suspects, to complement the new Houses of Parliament. The bridge was designed by Thomas Page in consultation with Sir Charles Barry. The contract was let to C. J. Mare of Millwall and work began in May 1854. However, in September 1855 Mare's business failed, delaying the work considerably. The bridge is of gothic design, harmonising with the Houses of Parliament, and has seven iron-ribbed spans. The central span is 130 ft, the third and fifth 125 ft, the second and sixth 115 ft and the end spans each 100 ft. The roadway is 58 ft wide with 13 ft footways on each side. The bridge was of structural interest as it was one of the first to use Robert Mallet's buckled metal plates patented in 1852, as the decking material. These have recently been replaced by reinforced concrete. After many complaints about the time taken, the bridge was completed by Cochrane & Co., and was opened on 24 May 1862.

Westminster Bridge. *The Engineer*, 19 September 1856, 511–12.

WALKER, R. J. B. *Old Westminster Bridge*. David & Charles, Newton Abbot, 1979.

YEOELL, D., BLAKELOCK, R. and MUNSON, S. R. The assessment, and strengthening of Westminster Bridge. *2nd Intl Conf. Bridge Management*, 1993, 307–15.

21. The Albert Embankment

HEW 2196
TQ 306 801 to
TQ 304 781

This embankment, on the south bank, runs from Westminster Bridge nearly to Vauxhall Bridge. The Act was obtained in 1863 and Bazalgette's Assistant Engineer, John Grant, supervised the work, which began in September 1865. The embankment was opened in May 1868. The contractor was William Webster. The reclaimed land behind the ¾ mile embankment provided a site for the building of St. Thomas' Hospital, which was re-sited as a result of the building of London Bridge Station.

22. Grosvenor (Victoria) Bridge

HEW 2253
TQ 287 777

The present Grosvenor Bridge was built in a phased construction programme during the years 1963–67. It replaced an original structure built in 1859–60 (then named Victoria Bridge) with later additions.

The 1859–60 bridge was the first rail bridge to cross the Thames in the London area. Carrying two mixed gauge tracks it provided access to Victoria Station (HEW 455, p. 173) for Great Western trains as well as those of the London Brighton & South Coast and the London Chatham & Dover companies. Sir John Fowler was the engineer;

Grosvenor (Victoria) Bridge

DENIS SMITH

the bridge comprised four 175 ft wrought-iron segmental arches over the river, with wrought-iron plate girder land spans of 65 ft at the south and 70 ft at the north end.

In 1865–66 a new bridge, designed by Sir Charles Fox & Son, also in wrought iron and repeating the elevation profile of Fowler's bridge, was added on the east side, enabling a total of seven tracks to be accommodated. The same form was again followed when further widening, this time on the west side, was carried out early in the twentieth century. Opened in 1907, this steel structure carried two more tracks, bringing the total to nine.

By the mid-twentieth century, with the older structures deteriorating and carrying heavier loads than they were designed for, complete replacement was appropriate. Aesthetic, technical and practical reasons influenced the decision to retain the four-arch profile. The new structure was built as ten separate single-track bridges, allowing staged replacement of the earlier spans with minimum interference to traffic. Each of the river spans consists of two two-pinned welded steel box-section arch ribs supporting a 13 ft 9 in. wide steel deck which carried the ballasted track. Substantial rebuilding of the piers resulted in the main spans being reduced to 164 ft. The land spans were replaced by welded steel plate girders.

Freeman Fox & Partners were the consulting engineers for the reconstruction, the main contractors being Marples, Ridgway & Partners. Dorman Long (Bridge & Engineering) Ltd. were the subcontractors for the fabrication and erection of the steel superstructure. The project was carried out under the general direction of Mr. A. H. Cantrell, Chief Civil Engineer, British Railways, Southern Region.

JACKSON, A. A. *London's Termini*. David & Charles, Newton Abbot, 2nd edn, 1985, 267–302.

TURNER, J. T. H. *The London Brighton & South Coast Railway*, 3 vols. Batsford, London, 1977, 1978, 1979.

MANN, F. A. W. *Railway Bridge Construction*. Hutchinson Educational, London, 1972, 30–5.

KERENSKY, O. A. and PARTRIDGE, F. A. The reconstruction of the Grosvenor Railway Bridge. *Proc. ICE*, 1967, **36**, 721–67.

WILSON, W. Description of the Victoria Bridge on the line of the Victoria Station and Pimlico Railway. *Min. Proc. Instn Civ. Engrs*, 1868, **27**, 55–120.

FOX, C. D. On the widening of the Victoria Bridge. *Min. Proc. Instn Civ. Engrs*, 1868, **27**, 68.

Waitrose
17 Liverpool Road
London
N1 0RW
Telephone no: 020 7276 2207

SALES VOUCHER
CUSTOMER COPY

Branch:	780	Date:	20/05/2009
Till:	054	Time:	19:19
Trans:	17581	Operator:	1156
Seq no:	5819		

Maestro UK ************5037
Type: CHIP
Merchant: 27330153 Auth code: 8051
AID: A0000000050001
Cryptogram: 40/C4A62561F98BA2A7
PAN Seq: 01 App Date: 081124 - 0210

GOODS £59.74

TOTAL £59.74

Please DEBIT my account with
the total amount shown

Cardholder PIN verified

Please retain this copy
for your records

www.waitrose.com
Food shops of the John Lewis Partnership

Doncastle Road, Bracknell Berkshire RG12 8YA
Waitrose Limited Registered Office
171 Victoria Street London SW1E 5NN
VAT No 232 457 280

Waitrose

23. Chelsea Suspension Bridge

HEW 2341
TQ 286 778

The present Chelsea Bridge, built 1934–37, is the second suspension bridge on the site. The first, opened in 1858 as a toll bridge, was designed by Thomas Page, who had earlier been Resident Engineer on the latter stages of the construction of the Thames Tunnel. Page's bridge was soon revealed as inadequate for London traffic and was strengthened in 1863–64, and further strengthened following its take-over by the Metropolitan Board of Works in 1879. By the 1920s its replacement was being seriously considered, but the financial crises of the period delayed action until 1933.

The replacement suspension bridge, designed by Ernest James Buckton and Harry John Fereday of Rendel Palmer & Tritton, was the first British suspension bridge of the self-anchored type to be designed. Taking advantage of the latest analytical techniques developed by American and Continental engineers over the previous 40 years it represented a major step forward in British bridge practice.

The foundation and piers of the new bridge, built in steel sheet-piled cofferdams, were located in a similar position to the previous structure, but were of completely new construction. The main suspension cables were made of 37 locked coil ropes bundled to form a hexagon. High tensile steel was used in the wires and in the flanges of the stiffening girders, one of the earliest applications, predating the first British Standard. The supporting towers were of steel box plate construction supported on rocker bearings on granite-encased concrete piles.

With a main span of 352 ft, centre to centre, and side spans of 173 ft, this was the most important suspension bridge built in Britain between the Wars. The contractors were Holloway Brothers, Furness Shipbuilding supplied the steelwork and the cables were by Wright's Ropes Ltd.

PAGE, G. C. On the construction of Chelsea Bridge. *Trans. Soc. Engrs*, 1863, 77.

BUCKTON, E. J. and FEREDAY, H. J. The reconstruction of Chelsea Bridge. *J. Instn Civ. Engrs*, 1937–38, **7**, 383–446; 1937–38, **9**, 429–46.

24. Chelsea Embankment

HEW 2196
TQ 285 779 to
TQ 274 777

This embankment, on the north bank of the river, runs for upwards of ¾ mile from Chelsea Hospital to Battersea Bridge. An Act was obtained in 1868, work began in July 1871, and it was completed in May 1874. This work reclaimed 9½ acres of foreshore land. The design and construction of this and the Albert Embankment are similar to, but without the services and transport complexities of, the Victoria Embankment. Between them the Victoria, Albert and Chelsea embankments extend over 3½ miles of riverbank and they made available about 52 acres of land.

25. Albert Bridge

HEW 205
TQ 274 775

The Albert Bridge is, in its hybrid cable-stayed form, perhaps the most unusual bridge across the Thames. Opened in 1873, it was designed by Rowland Mason Ordish, a gifted engineer who specialised in the design of iron structures; his work elsewhere in London included the domed roof of the Royal Albert Hall (HEW 300, p. 248).

When British engineers began to use iron cables to support bridges in the early nineteenth century they experimented with a variety of forms, including what could now be termed suspension bridges, with the load of the deck carried by hangers suspended from the main catenary chains, and cable-stayed bridges, with the deck suspended from inclined hangers. Albert Bridge represents a combination of both these systems.

Early experience of suspended bridges revealed the importance of a system of structural bracing to provide rigidity against movement induced by dynamic loads. In 1857 Ordish patented a system using a combination of stays and catenary cables to provide such rigidity. As modified for Albert Bridge the system comprises a main 'parabolic' cable designed to support the weight of the stays, a proportion of the dead and live load, through hangers at 20 ft intervals, together with the whole load of the central 40 ft section of the bridge. A series of inclined stays took the remainder of the load. Loads were calculated so that the design prevented any change in the proportion between the load supported at the apex of the main cable and that at each of its suspending points.

The bridge is founded on cylinder foundations supporting highly decorated iron towers. With a main span of 450 ft, the overall length is 711 ft. The principal cables were originally parallel wire steel rope, and the stays flat wrought-iron bars. The contractors for the ironwork were Andrew Handyside & Co. Originally a private toll bridge, following its takeover by the Metropolitan Board of Works it was strengthened by Sir Joseph Bazalgette in 1884 when the original wire cables, already rusting, were replaced by steel link chains. For most of its life the bridge had been subjected to a 5 ton load limit, and by the late 1960s this had been reduced to a 2 ton limit. During 1972–73 extensive strengthening repairs were carried out, which included replacing the bridge deck. Despite these works the overall appearance of the suspension of the bridge remains unaltered, although additional river piers and a transverse girder support the bridge at mid-span.

Greater London Council. *Report on the Condition and Proposed Remedial Works for Albert Bridge*. GLC, London, December, 1970.

MATHESON, E. *Works in Iron*, 1873, 170–175.

The Engineer, October & November, 1873, 281, 288, 301, 304, 316, 322.

Engineering, 1871, **11**, 373–74; 1872, **14**, 199–200.

26. Hammersmith Bridge

HEW 2248
TQ 230 781

The Hammersmith Company was established by an Act of Parliament of 1824 to construct the first suspension bridge over the Thames. On 22 February 1824 William Tierney Clark, who lived in Hammersmith as the Engineer to the West Middlesex Water Works Company, was appointed Engineer to the bridge project, and in March estimated the cost at £49 627. Clark's design was submitted to Thomas Telford who declared it 'highly satisfactory'. Tenders were invited and Captain Samuel Brown won the contract for the ironwork which was, however, made at the Gospel Oak Ironworks in Birmingham; G. W. & S. Bird of Hammersmith were awarded the stonework contract, and E. J. Lance of Lewisham that for decking and fencing. Work began in 1824 and the bridge was opened in October 1827. The north masonry pier was founded on 246 timber piles and the south on 256, the river clearance between the piers being 400 ft 3 in. There were eight chains of wrought-iron links, each having a cross-section of 5 in. by 1 in. Crowds watching the University Boat Race always produced unusual live loading, and in March 1869 the Company decided to test the strength, stability and state of repair of the bridge. They consulted R. M. Ordish and Col. W. Yolland, who decided to have some links tested at the Kirkaldy Testing Works in Southwark. These proved satisfactory and the bridge survived until the 1880s, when the Metropolitan Board of Works decided to replace the bridge to a design by Sir Joseph Bazalgette.

Bazalgette used W. T. Clark's river piers, so the present bridge is on the same alignment and spans as the previous bridge. His bridge has skeletal wrought-iron frameworks for the towers, which were clad in decorative iron castings. The chain links are of mild steel and were supplied by Sir John Brown of Sheffield at £33 per ton. Before acceptance, every link was proof loaded to 30 tonf/sq. in. at Kirkaldy's testing works in July 1886. Of the 844 links tested 100, were rejected leaving another 1228 to be delivered. There are four sets of chains, two on each side, with links 9 in. by 1⅛ in. and 9 in. by 1 in. The 6 in. diameter steel connecting pins were by the Thames Ironworks. The ornamental abutment and tower castings were supplied by the Whessoe Foundry, Darlington. The arches were manufactured by Thomson & Gilkes of Stockton. By early November 1886 the tower casings were fixed to the level of the cornice. W. T. Clark's old iron chains were erected to support a temporary wooden working platform constructed by John Mowlem & Co. By February 1887 the chains, suspension rods and stiffening girders were completed and the roadway was then constructed with Danzig Fir planks. The total cost of the bridge was about £71 500. It was opened by HRH the Prince of Wales on 18 June 1887.

SMITH, D. The works of William Tierney Clark. *Trans. Newcomen Soc.*, 1991–92, **63**, 181–207.

HAILSTONE, C. *Hammersmith Bridge*. Barnes & Mortlake History Society, London, 1987.

The new Hammersmith Bridge. *The Engineer*, 1887, **63**, 309, 330–31; 1887, **63**, 391–94.

27. Barnes Rail Bridge

HEW 353
TQ 214 763

The bridge was built to carry the Richmond and Hounslow loop line of the London & South Western Railway over the River Thames. Work began on this interesting structure in 1846 and it was opened on 22 August 1849. The bridge comprised three spans formed of cast-iron arches carrying a timber deck. The arches are of 120 ft span with a rise of 12 ft and the ribs are 3 ft deep. There are six arch ribs per span, each cast in four sections. The brick piers are faced in Bramley Fall stone. The engineers were Joseph Locke and J. E. Errington, and the contractor was Thomas Brassey. In 1891–95 the bridge was widened with wrought-iron bowstring trusses above the original cast-iron arches, giving the bridge an unusual appearance. The engineer for this work was Edward Andrews.

SZLUMPER, A. W. The reconstruction and widening of Barnes Bridge. *Min. Proc. Instn Civ. Engrs*, 1895–96, **124**, 309–22.

28. Chiswick Bridge

HEW 2249
TQ 203 763

In January 1925 the Ministry of Transport convened a conference between the Middlesex and Surrey authorities to discuss the proposed Great Chertsey (A316) arterial road scheme. This was designed to relieve Hammersmith Bridge and to avoid the congestion in Richmond. The project involved two new structures—the Chiswick and Twickenham bridges. On 3 July 1933 both bridges (and Hampton Court Bridge) were opened by HRH the Prince of Wales.

Built under an Act of 1928, Chiswick Bridge is a three-span reinforced-concrete structure with Portland stone facings. The central span is 150 ft and the other two are each 125 ft. Secondary stresses were avoided by inserting hydraulic jacks in the arch crowns. The carriageway is 40 ft wide with two 15 ft footways. The designer was Alfred Dryland, the Middlesex County Engineer, in association with Considere Constructions Ltd. Sir Herbert Baker was the architect. The contractors were the Cleveland Bridge and Engineering Co. and the cost was made up as follows: the bridge (including roadworks), £175 700; approaches, £32 700; land fees and supervision, £19 200, making a total of £227 600.

29. Kew Rail Bridge

HEW 2256
TQ 196 776

In 1864 the London & South Western Railway Company obtained an Act to extend their line from South Acton Junction to Richmond. Four pairs of cast-iron cylinders form piers supporting the three 115 ft long river spans of wrought-iron lattice girder trusses. There is a similar approach span at each end. Decorative cast-iron caps to the piers form the junctions between the girders giving a pleasing functional appearance to the structure. The bridge was designed by W. R. Galbraith and the contractors were Brassey & Ogilvie. The bridge was opened on 1 January 1869.

30. Brentford Dock and Augustus Way Bridge

HEW 2377
TQ 180 772

The Great Western, Brentford & Thames Junction Railway joined the Great Western Railway at Southall to a new dock at Brentford, for interchange with river barges serving waterside premises and the port of London. Enacted in 1855 the dock opened in 1859. I. K. Brunel was its

Brunel Bridge at Brentford

MALCOLM TUCKER

Chief Engineer, and William Davis Haskoll the 'Acting Engineer'. The Dock closed in 1964 and has been redeveloped for housing, around the retained dock basin. The quay walls are of composite brick and mass concrete construction with horizontal arches between counterfort piers, one of several economical designs of the 1850s. The river entrance has a large, single-leaf, tidal gate. A section of the original top beam, of wrought-iron box construction, has been preserved.

One of the original water-line girder bridges remains at Brentford, where the railway was converted in the 1970s to a road called *Augustus Way* (TQ 175 772). It crosses a riverside road, *The Ham*, on a 45° skew. The three wrought-iron plate girders, of 'half-through' configuration, have a 53 ft overall length and 5 ft 3 in. depth. Their details are characteristic of Brunel's work, with broad, light plates, shallow-headed rivets and (when originally built) widely spaced stiffeners. The bottom flanges are wider than the top flanges, which have a distinctive, segmental curved cross-section to increase the buckling resistance, with edge-stiffening angles additionally on the hog-backed girder.

The transverse deck beams were replaced, probably in the early twentieth century, with extra stiffeners and improved connection details. Precast, prestressed concrete slabs now support the roadway. The brick abutment walls make early use of recessed panels for better weight distribution—see the 'Three Bridges' (HEW 2269) further up the line.

JAMES, J. Russian iron bridges. *Trans. Newcomen Soc.*, 1982–83, **54**, 86.

31. Footbridge, Syon Park, Isleworth

HEW 2376
TQ 172 768

On the private land of the Duke of Northumberland's estate, this very early wrought-iron bridge carries a former driveway to Syon House across an ornamental lake. It isfrom a design of 1790 by the architect James Wyatt, with technical input probably from John Busch, lately returned from Russia. There are three spans, of 25 ft 6 in. and 30 ft 9 in. and 25 ft 6 in., but the deck follows one compound curve between the low stone abutments. Edge girders of flat plates, 12 in. deep, are stiffened by the parapets and strutted from beneath by very light arch ribs of

MALCOLM TUCKER

Footbridge, Syon Park, 1790

square-section, wrought-iron bar, with round hoops in the spandrels. The piers supporting them comprise two narrow posts, of bunched wrought-iron bars, connected by a transverse arched rib. The flat-plate deck incorporates secondary transverse and longitudinal ribs and is paved with gravel. The wrought- iron parapets are of rectangular bars, framed into St. Andrew's crosses, with subsidiary vertical railings, and they extend with a graceful flare beyond the abutments. There were formerly two similar bridges at Syon, of a single span. The elevation of the surviving bridge, but not the materials, resembles the cast-iron Metal Bridge on the Duke of Northumberland's estate at Alwick (HEW 1329), built substantially later in 1812.

32. Richmond Sluice and Half-Tide Lock

HEW 2195
TQ 170 751

The Thames at Twickenham had long given cause for concern to the navigation interests. In 1859 Richmond Vestry approached the new Thames Conservancy, without result. The Borough Surveyor said 'at low tide the stream was so shallow as to be insufficient to allow a pleasure steam-boat of the lightest draught to float along it, and its muddy foreshores were both unsightly and offensive'.[1] The problem was said to be due to the removal of Old London Bridge in 1833 and the increasing quantity of water being abstracted from the Thames by the London water companies above Teddington Lock. In 1871 Richmond Vestry again made an approach to the Thames Conservators requesting a lock and weir. The Conservators requested Sir John Coode and Captain Calver, RN, to report and also to confer with Sir James Abernethy, the engineer retained by the Vestry. Coode and Calver reported against the lock and weir but agreed that something should be done and proposed dredging. Little happened for ten years. In 1883 a Bill for a new lock and weir was promoted but this failed. However, in 1890 an Act authorising the work was obtained for the construction of a lock, a weir and a sluice, combined in an elegant footbridge.

The promoter of the work was The Thames Conservancy whose Engineer was C. J. More and the contractors for the bridge steelwork and the rotating sluice gates were Ransomes & Rapier of Ipswich. The designer of the sluices and steel arched bridge members was F. G. M. Stoney who, after a varied career as a consulting engineer, had become Manager of Ransomes & Rapier in 1887. The work comprises four concrete river piers faced

Richmond Sluice and Half-Tide Dock

DENIS SMITH

with Cornish granite and blue brick, two 6 ft wide footways, a clear 17 ft apart, with the sluices between. Three central spans of 66 ft each contain the Stoney Patent sluices, and two riverbank spans of 50 ft each span the lock (250 ft by 37 ft) on the Surrey side and the roller weir (Middlesex side). Each sluice gate is 12 ft high and weighs 32 tons and was originally operated by hand winch gear from the vertical to the horizontal position when opening, but is now electrified. The cost of the works was £61 000. The first pile was driven in July 1891 and the Duke of York opened the structure on 19 May 1894.

[1]BRIERLEY, J. H. Notes on the footbridge, lock and weir, Richmond, Surrey. *Proc. Inc. Assoc. Mun. County Engrs*, 1907–08, **34**, 133.

The Engineer, 25 May 1894, 439.

LOVEGROVE, E. J. Richmond footbridge and lock. *Proc. Inc. Assoc. Mun. County Engrs*, 1893–94, **20**, 354.

33. Twickenham Bridge

HEW 2250
TQ 172 748

This bridge was designed by Alfred Dryland with Maxwell Ayrton as architect. Work began on site on 1 June 1931. There are three river arches: the central span is 104 ft 4 in., two are 98 ft 4 in. Two land arches each

Twickenham Bridge

DENIS SMITH

span 56 ft. The three main arches are of a permanent three-pinned construction—the first use in a large reinforced-concrete bridge in Britain to be constructed on this principle. The pins at the arch springings and at the crown are masked by bronze coverplates, which somewhat detract from the clean lines of the arch profiles. The contractors were Aubrey Watson Ltd. and the total cost was £217 300. The bridge was opened on 3 July 1933.

34. Teddington Lock and Weir

HEW 2194
TQ 167 717

Due to shallows, this site had always been a source of trouble to the navigation. We know that in April 1775 some form of control was in place here as stops were put down 'to controul [*sic*] the current so as to form one navigable channel'. But the lock and weir, the lowest on the Thames, was the first structure built under the City of London's Thames Navigation Act of 1810 (5 Geo.III c.204). The lock was built 150 ft long and 20 ft wide. It was opened to traffic on 20 June 1811, and the accompanying weir was completed before the end of the same year. In 1825 the lock needed considerable repair, and in 1827 the weir was 'blown up' due to an accumulation of ice. By the 1840s the wash from steam traffic on the Thames was causing damage to banks and other structures. Following the removal of Old London Bridge it was found in 1848 that the water level on the lower sill of the lock had lowered by 2 ft 6 in. The lock was completely rebuilt in 1857–58, and the weir was rebuilt in 1871. This site has always been of importance as a boundary in subsequent legislation relating to the jurisdiction of the Thames.

35. Kingston Bridge

HEW 2347
TQ 173 694

Connecting Kingston and Hampton Wick, this bridge replaced a timber structure said to have been the first erected over the Thames after London Bridge. The present bridge was designed by Edward Lapidge, who was appointed County Surveyor of Surrey in 1824. The corporation of Kingston obtained an Act of Parliament in 1825. The Trustees applied to the Exchequer Loan Commissioners for funds and they submitted Lapidge's drawings to Thomas Telford for his approval, and having received this the Commissioners made a loan of £40 000. The bridge was built in masonry during 1825–58, and the Earl of Liverpool laid the first stone on 7 November 1825. The bridge has five river spans, the central arch of 60 ft the others of 56 ft and 52 ft, respectively, and two smaller arches, one on each side. The bridge was 27 ft wide. The Duchess of Clarence opened the bridge on 17 July 1828 and the bridge was freed of tolls on 12 March 1870 and was widened in 1914.

THORNE, J. *The Environs of London*. John Murray, London, 1876, 403.

BIRCH, J. B. B. An account of the bridge over the Thames, at Kingston, Surrey. *Min. Proc. Instn Civ. Engrs*, 1842, **2**, 184–86.

36. Chertsey Bridge

HEW 1834
TQ 054 666

The present bridge over the Thames replaces an earlier wooden structure first built about 1410. The new bridge, designed by James Paine of Chertsey, who also designed bridges at Kew and Richmond, was built by Charles Brown of Richmond between 1780 and 1785, and is sited just upstream of the site of the older structure.

The bridge, in Purbeck ashlar, has five main segmental arches, the centre span being 42 ft, with two adjacent spans of 36 ft each and outer side spans of 30 ft. In addition, on each side there is a 20 ft masonry brick arch over the towpath and two buried arches also of 20 ft span under the bridge approaches. The width of the bridge is 23 ft 6 in. between parapets. Originally the bridge had semi-circular recesses supported by brackets over the piers on either side, but these were removed in 1805 when they were found to be insecure. In 1820, when repairs were being carried out to the bridge, cast-iron grilles were inserted over the piers, but these were subsequently repositioned over the centre of each river arch during further major repairs carried out in 1991–92. The bridge is a scheduled Ancient Monument and Listed Grade II.

STRATTON, H. J. M. and PARDOE, B. F. The history of Chertsey Bridge. *Surrey Archaeological Collections*, **LXXIII**, 115–26.

LEACH, P. *James Paine*. Zwemmer, London, 1988, 179–80.

37. Staines Bridge

HEW 2089
TQ 031 715

This river crossing site must be the most ill-fated on the Thames. Since Roman times there had been a bridge over the Thames at Staines to carry the road from London to the West. Records date only from 1242 and this was for a wooden bridge which had suffered centuries of neglect, various inadequate grants of pontage and political expedients wholly inadequate to finance the proper maintenance of the structure.

Eventually, in 1791, an Act of Parliament authorised a new stone bridge to be built under the aegis of bridge commissioners, and which was designed by Thomas Sandby, then architect to the King's Works. Unfortunately, the foundations of the piers were too shallow and within a year of completion one pier settled so much that the whole bridge failed and had to be taken down. The commissioners were now understandably wary of pier instability and decided to trust the claims of Thomas Wilson, then a rising star in the latest construction material, cast iron, to build a single span from bank to bank. The bridge was built between 1801 and 1803, but within a short time the radial members started to crack due to

Staines Bridge

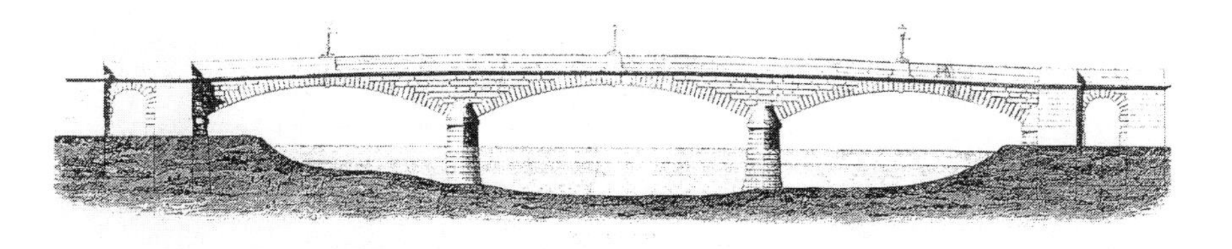

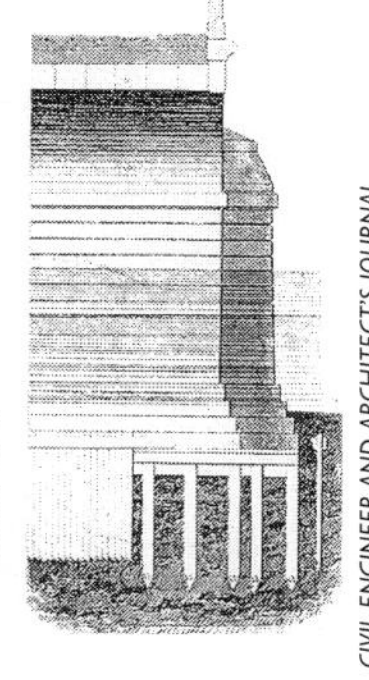

CIVIL ENGINEER AND ARCHITECT'S JOURNAL

outward movements of the abutments. John Rennie had been called in to advise, but in spite even of his efforts to arrest the movements the bridge continued to deteriorate and was taken down in 1806. Its replacement was largely wooden, but strengthened with iron plates. Parts of Wilson's bridge were used in the deck. This bridge lasted until 1827 when Rennie drew up outline plans for a stone bridge sited some 100 yd upstream above the confluence of the rivers Colne and Thames.

Designed by George Rennie, the new bridge was built of Aberdeen granite ashlar backed by Bramley Fall stone. It has three main segmental arches, the centre span of 74 ft and side spans of 66 ft. At each end of the bridge the return walls are pierced by 10 ft spans over the towpaths and there are further 20 ft span brick arches incorporated in the approaches, four on the north side and two on the south. The contractors, Jolliffe and Banks, started work in 1829 and the bridge was opened by William IV and Queen Adelaide on 23 April 1832. The total cost, including approach works, was £44 000. In 1958 the solid stone parapets over the main arches were removed and the bridge widened on either side by the provision of reinforced-concrete footpaths with metal railings. The original parapets were, however, retained at the ends of the bridge where the width between parapets was some 11 ft wider than over the main spans. The bridge is Listed Grade II.

PHILLIPS, G. *Thames Crossings: Bridges Tunnels and Ferries*. David & Charles, Newton Abbot, 1981, 134–39.

INSTITUTION OF CIVIL ENGINEERS. *Rennie Reports*, Vol. 4, 400 *et seq.*; *Rennie Collection*, GR 601, GR 719.

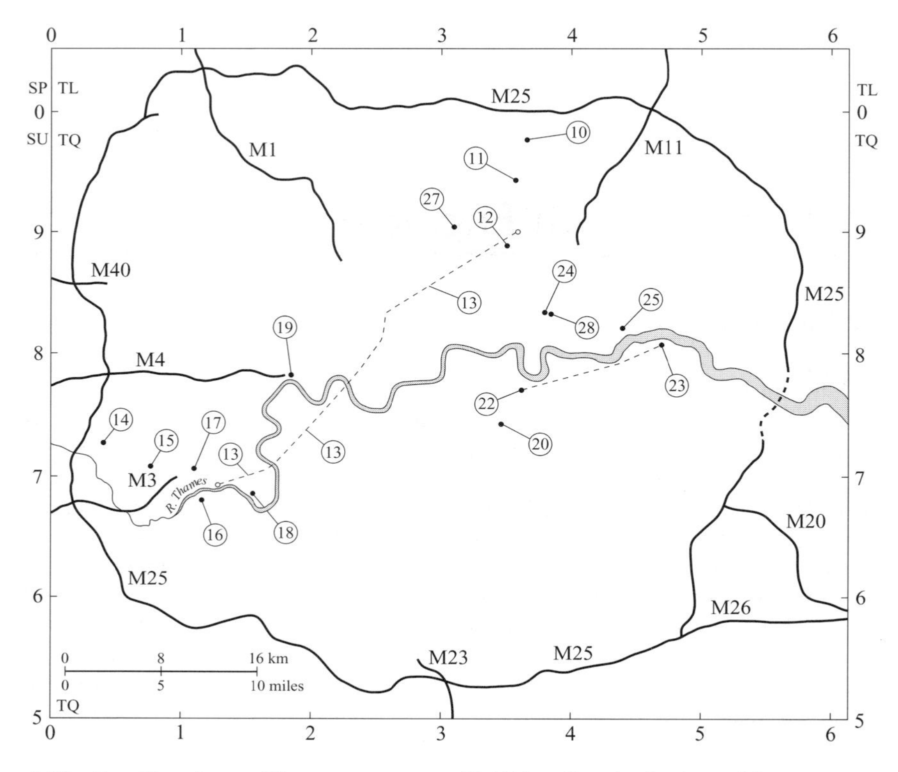

1. The New River (see p. 60)
2. New Gauge House (see p. 60)
3. Turnford Pumping Station (see p. 60)
4. New River Aqueduct over the M25 (see p. 60)
5. Cast Iron Aqueduct, Flash Lane, Enfield (see p. 60)
6. The Clarendon Arch (see p. 60)
7. Hornsey Treatment Works (see p. 60)
8. Stoke Newington Pumping Station, Green Lanes (see p. 60)
9. New River Head, Islington (see p. 60)
10. King George V Pumping Station and Reservoir, Enfield
11. William Girling Reservoir
12. Coppermill, Walthamstow Marsh
13. Thames–Lea Valley Aqueducts
14. Staines Reservoirs and Aqueduct
15. Queen Mary Reservoir
16. Walton Pumping Station and Reservoirs
17. Kempton Park Pumping Station
18. Water Supply to Richmond and Hampton Court Palaces
19. Kew Bridge Pumping Station
20. Honor Oak Reservoir
21. Thames Water Tunnel Ring Main
22. Deptford Pumping Station (see also p. 57)
23. Southern Outfall and Crossness Pumping Station (see also p. 57)
24. Abbey Mills Pumping Station (see also p. 57)
25. The Northern Outfall and Beckton Works (see also p. 57)
26. The Western Pumping Station (see also p. 57)
27. Markfield Road Pumping Station, Tottenham
28. West Ham Pumping Station

2. Public Health Engineering

Water Supply

Water has always been important as a means of transport, a source of energy, a type of defence and a boundary of land areas, but principally as the means of life support. By the middle of the nineteenth century, population growth and the expansion of urban living had posed serious health problems. The solution to these problems demanded the combined professional skills of the lawyer, physician, scientist, engineer and architect, together with the political will to finance vast engineering projects.

London's systems of water supply and main drainage had developed separately with different means of funding. Naturally, water supply was tackled first and attracted private capital—the early supplies were provided by companies which were more or less in competition with one another, at least at the supply-area boundaries. The public was prepared to pay these companies directly for a supply of drinking water. The history of water supply is therefore that of companies each serving relatively small areas and subsequently merging into ever larger units. This led to the formation of the Metropolitan Water Board in 1903 and subsequently to the Thames Water Authority in 1974. The public was not, however, so willing to pay for effluent removal. London had to await an appropriate local government structure, and adequate funds, before the appalling public health problems could be tackled effectively. These different origins have left their mark on the legacy of heritage sites and buildings.

Water undertakings comprise three main elements: capture, treatment and distribution. The first involves locating a water source and is a feature of even the most primitive of systems; treatment and widespread distribution require large capital investment and substantial engineering works. London is ideally situated for water supply, with its valley geology comprising a great chalk basin overlaid with gravel and clay. The earliest systems used natural springs with conduits (open channels or pipes) to convey the water, by gravity, to conveniently placed tanks or cisterns. Street names such as *White Conduit Street* and *Lamb's Conduit Street* are reminders of these schemes, as are the surviving hand pumps and well-houses. Wells were of two types; the common, or sunk, well was dug with pick and shovel, needed to be about 4 ft in diameter to accommodate the workmen, and was lined with either masonry or cast iron. The geology also made it possible to use a rotary auger to produce small-bore artesian wells near the level of the Thames to capture underground water. The first in London was bored in 1794 on Lord Holland's estate in Kensington. During the nineteenth century every large water consumer used them and there were artesian wells in breweries, hospitals and the naval dockyards. However, the principal sources for the water companies were, and still are, the Thames (71%) and the River Lea (14%), with wells and springs providing the remaining 15%.

The early water companies merely captured and distributed untreated water to their clients. However, in 1828, James Simpson, Engineer to both the Chelsea and Lambeth waterworks companies, built London's first slow sand filter bed at the Chelsea works. This filtration technique is still largely in use and the beds remain a distinctive feature in the waterworks landscape. By the middle of the nineteenth century, river pollution made it necessary to introduce stringent controls. The Metropolis Water Act of 1852 prohibited, from the end of August 1855, drinking water being taken from 'any part of the River Thames below Teddington' and required that 'Every reservoir within a distance in a straight line from Saint Paul's Cathedral in the City of London of not more than five miles ... shall be roofed in or otherwise covered over'. This was a landmark in water legislation and determined the location and type of many of the surviving structures and other aspects of civil engineering heritage.

Distribution of water over increasingly large areas during the eighteenth and nineteenth centuries required extensive pumping. London provides examples of every type of pumping technology, ranging from animal, water and wind power, Captain Savery's pumping engine, the Newcomen atmospheric engine, the reciprocating steam engine, gas, oil, steam and water turbines, the unique Humphrey pump, and the now ubiquitous electric motor. The Shadwell Water Company's seal on a document of 1785 clearly shows seven horses harnessed to the arms of a pump and the driver with his whip. Waterwheel-driven pumps were certainly in use in 1582 when the London Bridge Waterworks began to harness the tidal energy of the Thames. Waterwheels were progressively added to each of the first five arches from the City shore and remained until the bridge was demolished in the 1830s. In the late seventeenth century Savery's steam–vacuum engine was used to lift Thames water into tanks in York Buildings Waterworks, and in the eighteenth century Newcomen's atmospheric engines were employed by the New River and Chelsea companies. The East London Waterworks Company was the first to adapt and introduce the Cornish beam engine, after comparative trials in 1838, and, later, the triple-expansion steam engine into waterworks practice.

Most of the interesting surviving buildings and machinery date from the nineteenth century and were built for the private waterworks companies. By the middle of the nineteenth century the London water supply industry comprised nine companies with clearly defined supply areas. They were the New River (1609), Hampstead (1692), Chelsea (1723), Lambeth (1785), West Middlesex (1806), East London (1807), Kent (1809), Grand Junction (1811) and the Southwark and Vauxhall (1845). The only change in the company structure in the second half of the century occurred when the Hampstead Waterworks was taken over by the New River Company in 1856. The remaining eight companies survived until 1904 when they were replaced by the Metropolitan Water Board. The first Engineer appointed to the Metropolitan Water Board was W. B. Bryan, who had been Engineer to the East London Waterworks and he remained in post until his death in October 1914. The history of each company explains the location of the surviving works.

Two brief examples will serve to illustrate the complex relationships between engineering personnel, construction sites and an overall administrative body.

The West Middlesex Waterworks (HEW 2220) was established under an Act of 1806 to serve the parishes of Hammersmith, Kensington and part of Marylebone by drawing water from the Thames, storing it and pumping it on to the district. The company had difficulties with their early engineers. Ralph Dodd undertook the initial engineering work for the Act, but retired in 1806 before construction work began. At the beginning of 1807 William Nicholson began construction of the Hammersmith Works. Nicholson was replaced by Ralph Walker, who in turn resigned on 8 June 1810, and was replaced the same day by John Millington. William Tierney Clark was appointed Engineer on 23 January 1811 and remained so until his death in 1852. When Clark was appointed the company had just two reservoirs, one at Hammersmith storing 2.66 million gallons and another at Campden Hill which raised the total to 6.16 million gallons. Clark added Barrow Hill (Primrose Hill) in 1822, and Barn Elms in 1838, which brought the total storage capacity to 40 million gallons.

The Metropolitan Water Board was formed under The Metropolis Water Act of 1902 (2 Edw.7 c.41) and the new Board was to comprise a maximum of 69 members, of which the London County Council were to appoint ten and the other London boroughs 34. They met for the first time in April 1903 and the appointed day for taking over the undertakings of the water companies was 25 July 1904. The new Board obviously required its own Chief Engineer, and on advertising the post attracted 37 candidates. In June 1904 they reduced the list to three, namely W. B. Bryan (Engineer to the East London Company), E. Collins (Distribution Engineer to the New River Company) and H. F. Rutter (Engineer to the West Middlesex Company). W. B. Bryan was appointed at a salary of £2500 and had to give up his private consultancy practice. Bryan was succeeded by Sir James Restler, who was in post from 1914 to 1918, and H. F. Cronin was appointed Chief Engineer in 1939 and superintended the Board's works during the Second World War, and beyond. They took over the former eight companies and re-zoned the area, although some of the old company boundaries were retained. But the old division between water supply and main drainage remained. It was not until the 1970s that the Thames Water Authority, a body with control over the whole water cycle, was formed.

Main Drainage

In the case of main drainage public health was seen to be a matter for public wealth. This meant that there was a strong political dimension to achieving an effective system of main drainage, both in terms of financing any large-scale project and also of establishing an appropriate local government authority for London with capital-wide powers. A series of disastrous epidemics of cholera from the 1830s brought public health issues to the forefront of public debate. The individual parishes and vestries could only tackle their problems in a piecemeal manner, and by the 1840s central government intervention became inescapable, leading to the Government establishing the first Metropolitan Commissioners of Sewers in 1848. Each *ad hoc* Commission comprised 12 members and there were six Commissions between 1848 and 1855. In 1849 Joseph William Bazalgette was appointed 'Assistant Surveyor' to the second Commission and thus began his 40-year career in local government engineering. The City Corporation had their own Commissioners of Sewers under the City Surveyor (later Engineer) William Haywood. Both the Metropolitan and City Commissioners occasionally employed consulting civil engineers, including I. K. Brunel, Robert Stephenson, Sir William Cubitt, Thomas Hawksley and George Parker Bidder. In 1852 Bazalgette was appointed Engineer to the Metropolitan Commission, and he remained as Engineer to the last two Commissions. During this period the main drainage scheme of separate interceptor sewer systems north and south of the Thames had evolved. In 1855 the Government established the Metropolitan Board of Works with powers over 117 square miles of the Capital. They took power on 1 January 1856 and Bazalgette was elected Engineer to the Board in the same month. Bazalgette developed and implemented the main drainage scheme for London built between 1859 and 1874. As a local government

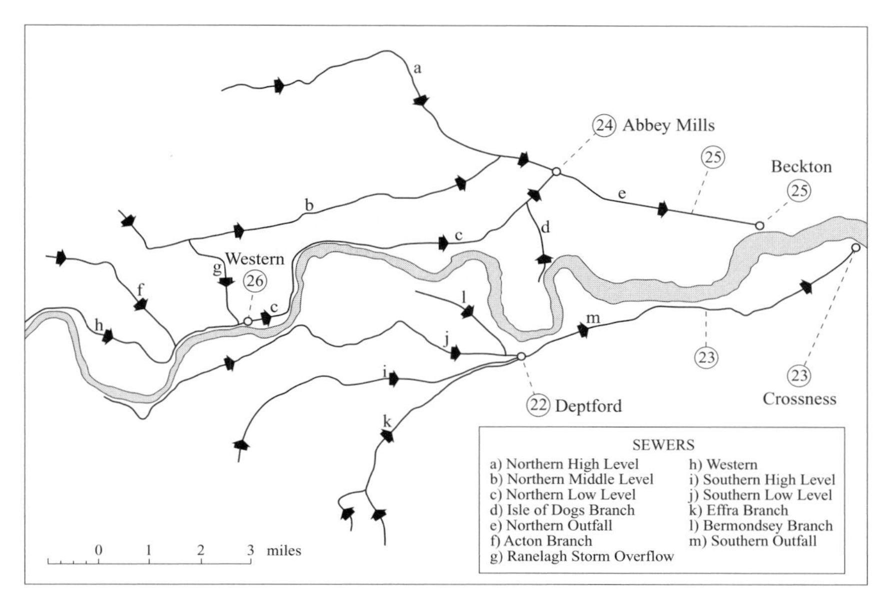

Metropolitan Board of Works main drainage system

engineer Bazalgette not only designed and supervised the construction of the works but, once completed, ran and maintained the system during the 33 years of the Metropolitan Board of Works.

The Southern system comprised interceptor sewers running west to east with a lift pumping station at Deptford and a Southern Outfall sewer to Crossness pumping station. The southern system was operational in 1865. The northern main drainage system comprised three levels of interceptor sewers running west to east, each falling at 2 ft per mile and at high, mid- and low levels. The northern pumping stations are at Pimlico and at Abbey Mills in east London, with an outfall sewer from Abbey Mills to Beckton, where the sewage was allowed to settle before discharge into the Thames. The northern main drainage took longer to complete because of the difficulties involved in incorporating the low-level sewer and the Metropolitan District Railway into the Victoria Embankment. Abbey Mills pumping station was opened in 1868 and the whole London system could be regarded as complete on the opening of the Western Pumping Station at Pimlico in 1875.

SMITH, D. (ed.). *Water Supply and Public Health Engineering*. Ashgate, Aldershot, 1999.

SMITH, D. The works of William Tierney Clark, *Trans. Newcomen Soc.*, 1991–92, **63**, 181–207.

DICKINSON, H. W. *Water Supply of Greater London*. Newcomen Society, London, 1954.

BERRY, G. C. *London's Water Supply 1903–1953: A Review of the Works of the Metropolitan Water Board*. Metropolitan Water Board, London, 1953.

SISLEY, R. *The London Water Supply: A Retrospect and a Survey*. Scientific Press, London, 1899.

BAZALGETTE, SIR J. On the main drainage of London, and the interception of the sewage from the River Thames. *Min. Proc. Instn Civ. Engrs*, 1865, **24**, 280–358.

Water Supply Heritage in the Lea Valley

1. The New River

HEW 277
TL 350 137 to
TQ 314 827

This water supply system was initiated by the Corporation of the City of London, which obtained an Act, in 1606, to bring fresh drinking water, by gravity, from chalk springs in Chadwell and Amwell in Hertfordshire to Islington in the heart of the City. The Corporation engaged the entrepreneur Sir Hugh Myddleton to finance and implement this scheme, which was built between 1609 and 1613. The water supply from the springs at Chadwell and Amwell was supplemented from 1700 by water from the River Lea below Hertford and in the nineteenth century deep wells were sunk alongside the route to obtain extra water from the chalk. The New River was opened, with due ceremony, when water was admitted to the Round Pond at New River Head in Islington on 29 September 1613. The New River Company was well served by a father and son succession as Surveyors for 90 years. Robert Mylne was appointed Joint Surveyor in November 1767, becoming Surveyor in July 1771. On his death in 1811 he was succeeded by his son, William Chadwell Mylne, who remained in the post until 1861.

The river was originally 40 miles long, following the 100 ft contour on the west side of the Lea Valley, but its many loops were straightened in the 1850s, saving 12 miles. From the terminal reservoir at New River Head in Islington, water was originally conveyed to the City, and later to the West End also, in pipes bored from elm trunks. These were replaced in cast iron from 1810. To meet new legal requirements and expanding demand, waterworks with filter beds were built after 1852 at New River Head, Stoke Newington and Hornsey. In 1946 the Works at New River Head were closed, so that the New River terminated at Stoke Newington. The Historical Engineering Works of The New River Company are described geographically, from north to south.

BERRY, G. C. Sir Hugh Myddleton and the New River. In: Smith, D. (ed.), *Water Supply and Public Health Engineering*. Ashgate, Aldershot, 1999, 46–78.

GOUGH, J. W. *Sir Hugh Myddleton: Entrepreneur and Engineer*. OUP, Oxford, 1964.

The New River

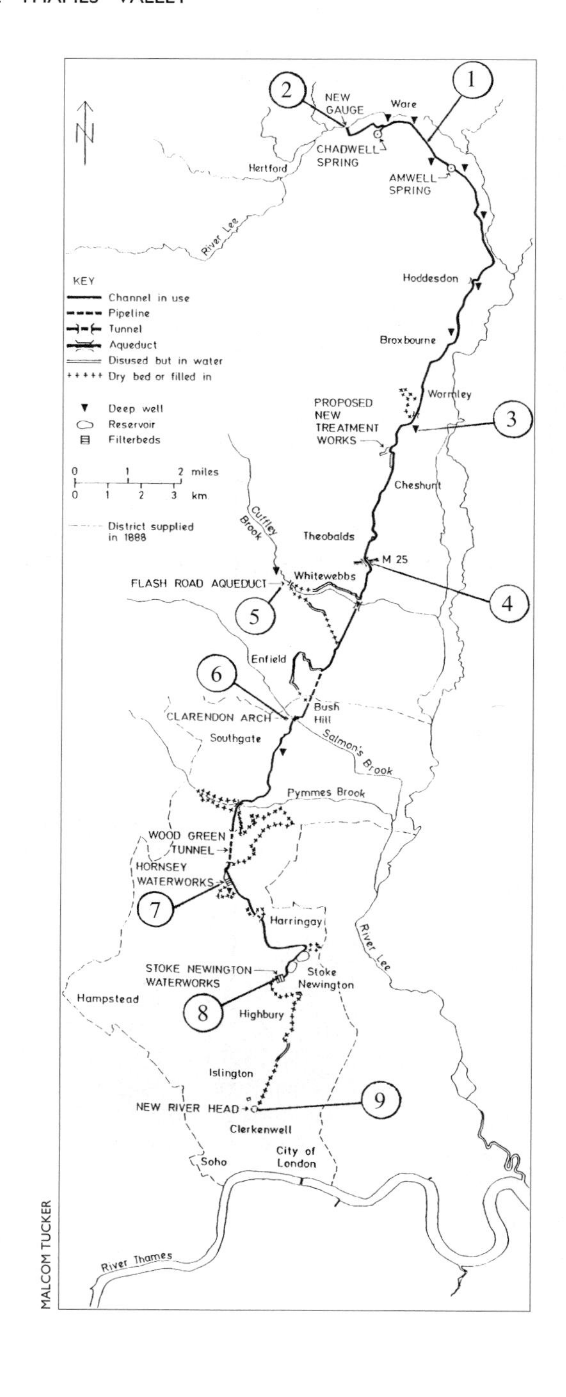

DENIS SMITH

New River, cast-iron bridge, 1824

DENIS SMITH

New River, Amwell Pond

2. New Gauge House

HEW 277
TQ 340 138

By the end of the seventeenth century the New River Company obtained powers to augment their supply from the two springs by abstracting water from the River Lea. This was monitored by a wooden 'balance engine', a rocking beam-and-float device following the rise and fall of water in the Lea. This device was replaced by the surviving 'Marble Gauge' in 1770. In 1855 the Company obtained another Act, enabling them to abstract 2500 cu. ft per minute from the Lea, and this led to the design and construction of the present *New Gauge House*, built in stock brick with a slate roof immediately adjacent to the river. The gauge itself comprises two floating rectangular pontoons connected by a wrought-iron bowstring girder. Between the pontoons a metal sluice, 7 ft 10 in. wide and suspended from the girder, spans the channel bringing water from the Lea. This rises and falls with the level changes in the Lea and maintains a constant depth of water over the sill of the sluice of 16 in. The gauge maintains a steady flow of water at 2.72 miles per hour, delivering 22.464 million gallons per day into the New River.

New Gauge House

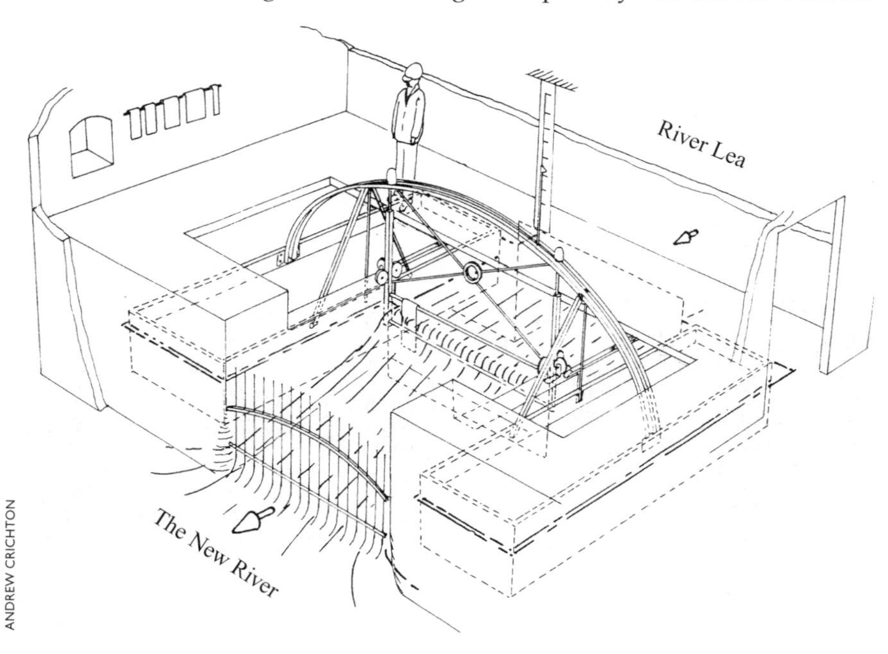

ANDREW CRICHTON

3. Turnford Pumping Station

HEW 1925
TQ 360 045

This is one of the well-pumping stations built alongside the New River in the nineteenth century to augment the supply from the springs and the River Lea by abstracting water from the chalk and discharge into the New River. The most northerly of these pumping stations is Broadmead Pumping Station (TQ 356 140) at Ware.

The Turnford buildings are, unusually, in red brick with stone dressings. The two principal buildings, erected in 1870, are the engine house with a shallow-pitch slate roof, and the tower over the well with lifting tackle and of sufficient height to lift out the pump rods. The Turnford engine house is the last one on the New River to contain a steam engine. The engine, by Boulton & Watt (although not marked), is of the side-lever marine type with a 15 ft diameter flywheel with a connecting rod ('pitman') driving a bell-crank lever over the 180 ft deep well. This engine was originally installed in 1848 at the Hampstead Road well, begun in March 1835, of the New River Company.[1] The well-pumping gear was supplied by Hunter & English of Bow. In August 1882 a second steam engine, by Moreland & Sons, was installed on its

New River, Broadmead Pumping Station

WENDIE TEPPETT

own girder over the well, but was subsequently removed when electrification took place in 1953.

[1]MYLNE, R. On the supply of water from artesian wells in the London basin, with an account of the sinking of the well at the reservoir of the New River Company, in the Hampstead Road. *Trans. ICE*, 1842, **3**, 229–44.

4. New River Aqueduct over the M25

HEW 2214
TQ 343 001

The seventeenth-century New River, running north–south, proved an interesting challenge facing engineers constructing the M25. The New River at this point (just west of the Enfield Tunnel) crosses the path of the motorway on high clay embankments and an aqueduct was therefore necessary. The Department of Transport's Agent was the Greater London Council, who decided on a post-tensioned, cast *in situ*, concrete structure carrying the New River in two rectangular boxes 90 m across the twin three-lane carriageways running east–west. The double-box design allowed the temporary diversion of the river during construction to be done in two stages. Steel sheet piling was used to redirect the river. The concrete box elements are each 4.25 m wide and 2 m deep, giving an overall width of 10.5 m. The top slab of the aqueduct boxes enables Thames Water to use the structure as an access road. The double-box design allows subsequent maintenance work to be done on one side at a time without interrupting the vital water supply of 200 million litres per day. The concrete boxes are lined with epoxy panels to prevent contamination. The aqueduct was completed in 1985. The contractor was Sir Alfred McAlpine.

New Civil Engineer, 5 September 1985, supplement, 42.

5. Cast Iron Aqueduct, Flash Lane, Enfield

HEW 277
TQ 323 994

This structure resulted from improvements made to his estate by the owner of *Claysmore* in the early nineteenth century. Having bought the loop of the New River he converted the area within the loop into a lake by damming the Cuffley Brook. The New River was diverted, and

DENIS SMITH

Cast Iron Aqueduct, Flash Lane, Enfield

in 1820–21 the cast-iron aqueduct and embankment was constructed to carry the New River over Cuffley Brook. The two-span cast-iron trough, carried on brick piers and segmental arches, has a base plate with five integral fish-bellied cast-iron upstand stiffeners about 1 in. thick. The ironwork was cast by Hunter & English of Bow, at a cost of £252 2s. The Enfield Archaeological Society excavated the site and made measured drawings of the aqueduct in 1970, and the ironwork is clearly visible and accessible from Flash Lane.

Enfield Archaeological Society. *Industrial Archaeology in Enfield*. Enfield Archaeological Society, London, 1971, Research Report No. 2.

6. The Clarendon Arch

HEW 277
TQ 325 951

This brick tunnel culvert carries the New River over Salmons Brook, through an embankment, and is the oldest surviving structure on the New River. The brick archway replaced the original timber-boarded aqueduct. The brick arch, dated 1682, carries the arms of the Earl of Clarendon (Governor of the New River Company). The western entrance carries the inscription 'This Arch was rebuilt in the Yeare 1682, Honorable Henry Earle of Clarendon on being Gov.'. The arch was

DENIS SMITH

Clarendon Arch under renovation

badly cracked and had timber shoring to support it for many years, although Thames Water have recently carried out restoration work on the structure, which is Listed Grade II.

7. Hornsey Treatment Works

HEW 2215
TQ 306 897

The New River arrives at these works through the 1100 yd long 14 ft diameter Wood Green tunnel built by W. C. Mylne in 1852 to replace a meandering part of the original contoured river course. Leaving the tunnel the river passes under an arch of the Great Northern Railway into the Hornsey Basin with an area of 1 acre. The water then passes southwards through filter beds. A red-brick Sluice House straddles the river. The main pumping station on this site is of red brick with stone dressings, and is dated 1903, making it the last building to be constructed by the New River Company before the Metropolitan Water Board took over the operation in 1904. The red-brick pumping station survives and still contains oil engines by W. H. Allen of Bedford. Nearby is the Campsbourne Well.

8. Stoke Newington Pumping Station, Green Lanes

HEW 2216
TQ 323 869

Under an Act of 1830 the New River Company leased 50 acres to build two reservoirs here—the East and West. Both reservoirs were completed and filled by the spring of 1833. The Engineer was William Chadwell Mylne. An Act of 1852 enabled the reservoirs to be deepened and two filter beds were constructed. Mylne also designed the exotic 'castle' pumping station in 'Scottish Baronial' style which was completed in 1855. The original pumping station comprised six beam engines, two single-cylinder engines by James Watt & Co., and four compound beam engines by James Simpson. An interesting feature of the architecture was that the engine flywheels projected into the hollow external buttresses. The building, Listed Grade II*, has been converted into an indoor climbing training facility without affecting the exterior elevations.

Illustrated London News, 1852.

Stoke Newington Pumping Station

WENDIE TEPPETT

9. New River Head, Islington

HEW 2217
TQ 313 827

This was the London terminus of the New River in Islington, and is one of the most important engineering heritage sites in London. Originally, water from here gravitated from the Round Pond to serve the district below. As London developed it became necessary to back-pump the water to a higher reservoir constructed on Islington Hill. New River Head is of great interest, in engineering history terms, in that to undertake this pumping, horse-engine pumps, a windmill, a Newcomen atmospheric engine, a waterwheel, Boulton & Watt steam engines, a triple-expansion steam engine and electric motors have all been employed at various times.

In 1708 the New River Company leased an acre of land in Islington from the Earl of Clarendon to build the 'Upper Pond', later to be known as Claremont Square Reservoir. Shortly before this the Company had consulted George Sorocold of Derby, who proposed the upper reservoir, a windmill to pump from the round pond to the upper pond, and a horse-engine to maintain supply during calm weather. The lower part of the tower mill remains. In 1766 John Smeaton reported suggesting a Newcomen-type engine. This was built to John Smeaton's design; it was his first beam engine and was in operation in 1768. In 1818 two wing buildings were added to Smeaton's building to accommodate two Boulton & Watt steam engines and these engine houses survive. Between 1897 and 1903 the Boulton & Watt engines were replaced by two triple-expansion steam engines, the first by James Simpson & Co. and the other by Yates & Thom of Blackburn. They last worked in 1950.

In the adjacent Metropolitan Water Board building of 1920 is incorporated the original Board Room of the New River Company. It is the 'Oak Room' of 1693, with an exquisite plastered ceiling, containing the Arms of Sir Hugh Myddleton, and wall panels with watery themes carved by Grinling Gibbons.

10. King George V Pumping Station and Reservoir, Enfield

HEW 2222
TQ 373 979

This pumping station is of international importance as being the site of the world prototype installation of Herbert A. Humphrey's unique pumps. The Humphrey pump is a device without a piston, connecting rod, crank, flywheel or gearing, and is one of the most thermally efficient prime movers yet invented. Humphrey patented it in 1906, and W. B. Bryan, Chief Engineer to the Metropolitan Water Board, was present at a lecture given by Humphrey in London in 1909. At that time Bryan was considering the options available to him for lifting water from the River Lea into the large reservoir then under construction. He bravely decided on the Humphrey pump after seeing a small test-bed demonstration. The pump is basically a large U-tube with a combustion chamber at the closed end and an open delivery outlet at the other. The water surface in the combustion head forms the piston, and water oscillates in the tube operating on a four-stroke cycle, where each stroke is of a different length. The periodicity of the pump depends on the diameter and length of the pipe. Five pumps were ordered, four to lift 40 million gallons per day each and the fifth of 20 million gallons per day capacity. They were

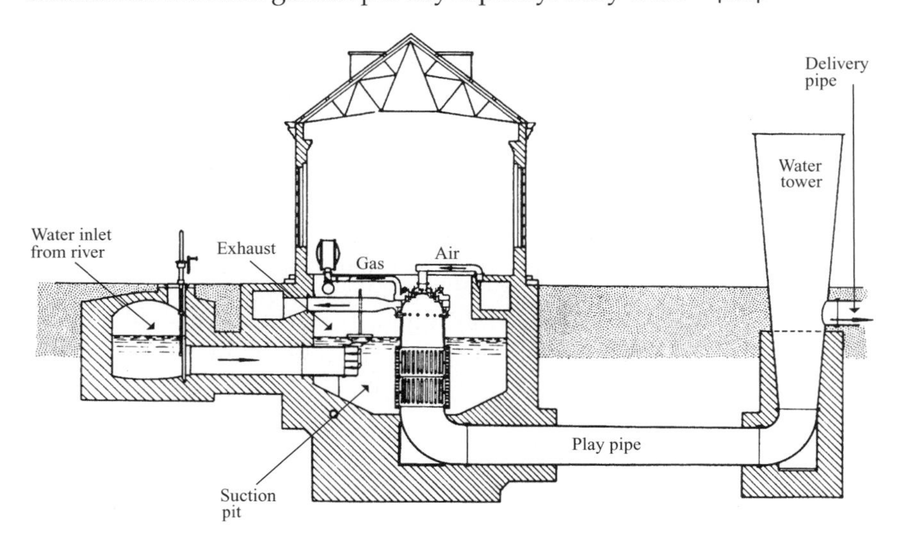

King George V Pumping Station, the Humphrey pump

built by Siemens Brothers of Stafford. The fuel was gas supplied from a Dowson producer gas plant with a small gasholder. The pumping station worked on an intermittent basis and was operated when there was a large quantity of water flowing down the Lea which, if not stored in the reservoir, would cause flooding lower down the valley. The raw water then flowed down an open aqueduct for subsequent treatment. The pumping station was opened by King George V on Saturday 13 March 1913 and received world-wide press coverage. Although interest in the pump's principle was international its arrival was ill timed. After the First World War electricity generation and distribution made rapid strides and the device fell out of favour. In the 1960s an attempt was made to convert the pumps to use natural gas, but by this time the need for the station had declined. Of the five original pumps only three survive *in situ*, their survival is aided by the fact that most of a Humphrey pump is buried underground.

SMITH, D. The Humphrey pump and its inventor. *Trans. Newcomen Soc.*, 1970–71, **43**, 67–92.

11. William Girling Reservoir

HEW 2221
TQ 366 940

The reservoir is bounded on its northern side by *Lea Valley Road* (A110). This was another construction project interrupted by the Second World War. Work began in 1935 when the tender of John Mowlem (for £682 156) was accepted in July. This work, designed by Sir Jonathan Davidson, attracted widespread technical interest in 1937 when a major slip occurred in the partly formed embankment at the north-west corner when the embankment fill had reached a height of about 23 ft. A 66 ft width dropped 2 ft 4 in. and moved forward 13 ft. Fortunately, the dam failed before any water was stored. Investigations were underway when a second slip occurred in December. Two independent soil mechanics experts were called in, Dr. Herbert Chatley and Professor Karl Terzaghi, who both made recommendations. In July 1938 the Metropolitan Water Board approved important modifications to the original design. Subsequent investigations into this landslip can be regarded as the birth of modern soil mechanics in Britain. The reservoir was redesigned, to increase its capacity by 11.3%. As completed the reservoir has a perimeter of 3.5 miles, a water area of 334 acres and a capacity of 3500 million gallons.Originally known as the Lee Valley Reservoir at Chingford, it was renamed, after the Chairman of the Metropolitan Water Board, on its opening on 4 September 1951.

GOODMAN, R. E. *Karl Terzaghi: The Engineer as Artist*. ASCE Press, Reston, VA, 1998, 175–76.

12. Coppermill, Walthamstow Marsh

HEW 2218
TQ 351 882

There has been a mill on this site for centuries. It had been a corn mill, and gunpowder was made on this site in the seventeenth century—a map of 1699 describes the area as *Powdermill Marsh*. From about 1690 it was a paper mill for about 15 years. By 1710 it was a leather mill, and from 1742 was used as an oil mill—crushing flax to extract linseed oil.

The British Copper Company was formed in 1807 and purchased the mill in 1808. The Company smelted copper at Landore, near Swansea in south Wales, and the ingots were brought by sea, up the Thames and the Lea

Coppermill, Walthamstow Marsh

DENIS SMITH

Navigation to the Walthamstow mill where they were rolled into sheets for general use and also for stamping into coinage tokens. The rolling mill machinery was installed by Lloyd and Ostell in 1809. At that time the undershot waterwheel was 18 ft diameter and 20 ft wide, ran at 5.7 rev./min and drove a pair of 16 in. diameter rollers through bevel gearing.

The rolling mill ceased working in 1857 and the machinery was transferred to Swansea. The building was then bought by the East London Waterworks Company who used the waterwheel to pump water. The East London Company also added the Romanesque tower on the west side of the building to accommodate a Cornish 'Bull' engine. On the south side of the building is a triangular peninsula on which a hand crane with a wooden jib survives. The building, which is in stock brick with a pantiled hipped gable roof, passed into the hands of the Metropolitan Water Board and is now owned by the Thames Water Authority who maintain the building in good condition for use as a store. Opposite is the modern Coppermill Water Treatment Works of the Thames Water Authority. The building is easily seen from *Coppermill Lane*.

Water Supply Heritage in West London

13. Thames–Lea Valley Aqueducts

HEW 2234

These aqueducts are the links between the west London reservoirs and pumping stations and the Lea valley.

With the removal of the abstraction sites from inner London in the 1850s to west London there arose the need to transport filtered Thames water taken from above Teddington Lock to serve districts in north London. Various projects were undertaken to achieve this object. These included large cast-iron underground mains constructed in the late nineteenth century and, more recently, by means of a concrete-lined tunnel. The idea was first suggested in 1935 but, again, the War intervened and the scheme was postponed. In 1949 an experimental 1000 ft. long tunnel was begun at Stoke Newington waterworks and completed in 1951. Boring in London

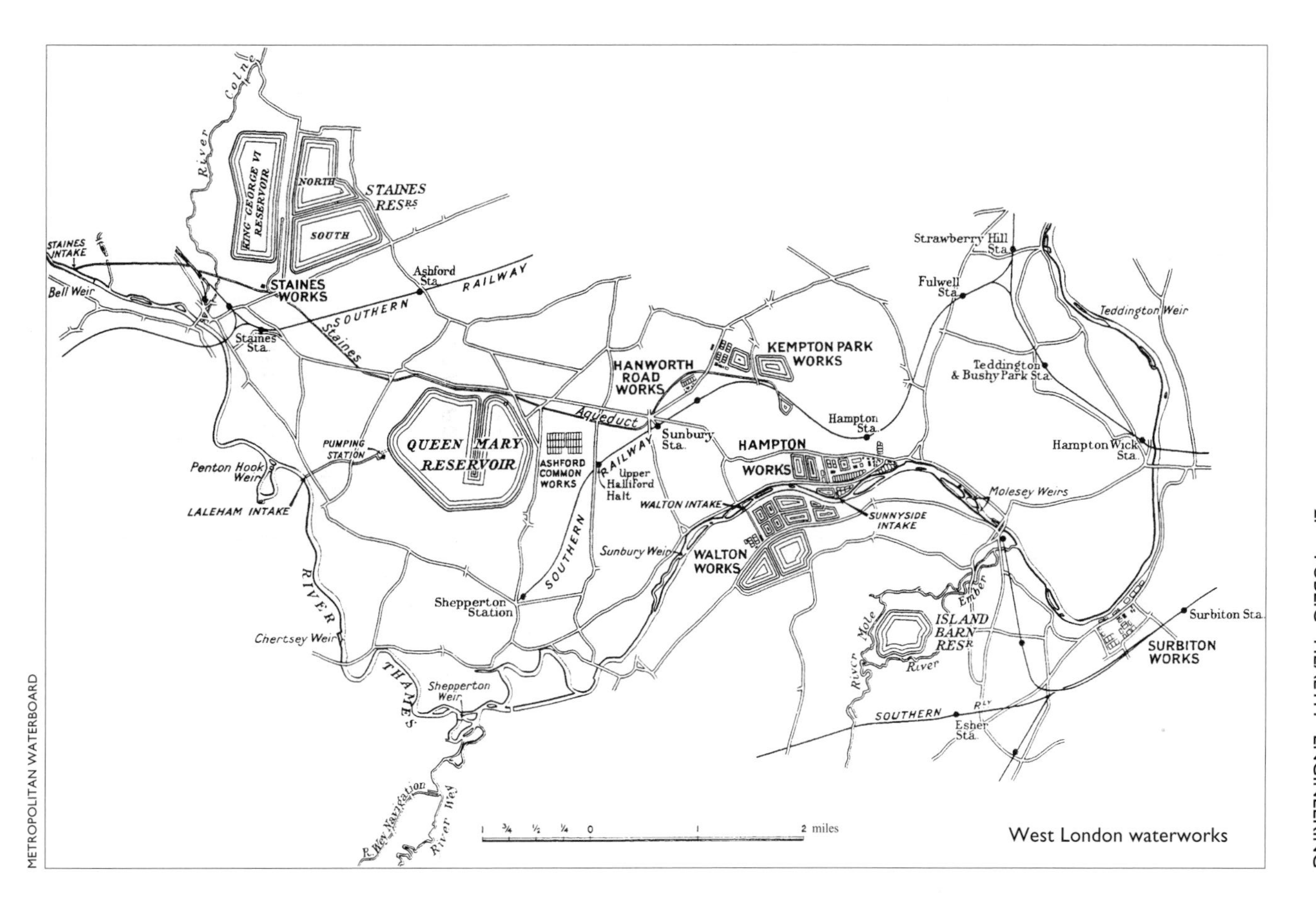

West London waterworks

METROPOLITAN WATERBOARD

clay presented little difficulty and the 90 in. diameter tunnel was lined with interlocking concrete rings. In the final design the tunnel diameter was increased to 102 in. Work on the remainder of the 19-mile route started at the Hampton Works, and was taken under the river near Teddington Lock. It continued under Ham Common, Richmond Royal Park, Barnes Common and crosses the river again below Hammersmith Bridge. It then passes close to Olympia, Paddington Station, and Lords cricket ground, Holloway Prison and on to Stoke Newington and finishes at the Lockwood Reservoir in the Lea Valley. The scheme was designed to re-stock the Lea Valley reservoirs in times of water shortage. The design and supervision of civil engineering works was by Sir William Halcrow and Partners and the principal tunnelling contractors were Kinnear, Moodie & Company, A. Waddington & Son, and Balfour Beatty. The work cost £4 300 000 and was inaugurated in September 1960.

14. Staines Reservoirs and Aqueduct

HEW 2225

The development of this site was undertaken by the Staines Reservoirs Joint Committee, comprising the New River, Grand Junction and West Middlesex Waterworks Companies. In 1896 they obtained an Act (59 & 60 Vict. c.241) to build two reservoirs, the Staines North (TQ 050 735) and South (TQ 054 727). In July 1937 the Metropolitan Water Board let a contract to John Mowlem for a third reservoir on the west side of the earlier reservoirs. Work began in August 1937, but it was not completed until after the Second World War. The new work, called the King George VI Reservoir (TQ 042 735), occupies 350 acres, holds 3493 million gallons and was opened in November 1947. These reservoirs are supplied by the Thames from the Staines intake just above Bell Weir, and the *Staines Aqueduct* then continues eastwards to supply the Kempton Park and Hampton works.

15. Queen Mary Reservoir

HEW 2226
TQ 073 695

Lying just south of the A308, this was the first large reservoir completed by the Metropolitan Water Board. S. Pearson & Son Ltd., the contractors, completed their work in December 1924. The reservoir was built to hold 6700 million gallons with a top water area of 707 acres. King George V officially opened the reservoir in June 1925. The reservoir, originally referred to as the 'Littleton Reservoir' was renamed on the opening day. It was the largest reservoir in the world when built and remained so for at least a quarter of a century. This reservoir is supplied from the Thames from the Laleham intake. The reservoir embankment was damaged by bombs during the Second World War, but as the Metropolitan Water Board had taken the precaution of lowering all reservoir water levels by 5 ft at the at outset of the War no water was lost at this or any other similarly damaged reservoir.

16. Walton Pumping Station and Reservoirs

HEW 2229
TQ 119 684

These works, built by the Metropolitan Water Board, comprise two reservoirs, an intake from the Thames, filter beds and a pumping station. The works were empowered by an Act of 1898 obtained by the Southwark and Vauxhall Waterworks Company, but work did not start for some years. The *Knight* (western) reservoir has a capacity of 480 million gallons with a water area of 51.5 acres, and the *Bessborough* (eastern) reservoir contains 718 million gallons and a water area of 74 acres. The works comprised 1.5 million cu. yd of excavation, provision of 200 000 cu. yd of puddle clay and 75 000 cu. yd of Portland cement concrete. The works cost approximately £419 000 and were designed and supervised by J. W. Restler, former Engineer to the Southwark & Vauxhall Company, who had been appointed Deputy Chief Engineer to the Metropolitan Water Board in 1905. The reservoirs were inaugurated on 13 April 1907. The pumping station began working on 10 June 1911. The machinery comprised four triple-expansion steam engines with an aggregate of 2000 hp. The engines were supplied by The Thames Ironworks Shipbuilding and Engineering Co. of Greenwich. Each engine drove a centrifugal pump capable of lifting 25 million gallons of water per day into the reservoirs, which then flowed by gravity to the Hampton works. One of the steam engines survives in the pumping station and is maintained in excellent condition by Thames Water.

17. Kempton Park Pumping Station

HEW 2231
TQ 111 706

The original works on this site were planned by the New River Company, but were soon taken over by the Metropolitan Water Board. The construction period was 1902–05. A stock brick pumping station, boiler house, chimney, and 12 slow sand filter beds were built. The pumping machinery comprised five triple-expansion steam engines supplied by the Lilleshall Co. Each engine developed 100 hp, from steam at 150 lbf/sq. in. and could

Kempton Park, the North Engine

KEMPTON GREAT ENGINES TRUST

deliver 15 million gallons per day. The engines were removed in the 1960s but the structure remains and is still known as the 'Lilleshall Building'. In the 1920s Henry Stilgoe, the Metropolitan Water Board Chief Engineer, produced plans for a new Engine House, close to the Lilleshall Building, and a new Primary Filtration House on the other side of Sunbury Road. The new building is of red brick with stone dressings. The contractors were William Moss & Sons Ltd. Two of the largest triple-expansion steam engines ever built for water supply were installed, each of 1008 hp and cost £94 000. Six Babcock & Wilcox water-tube boilers provided steam, together with a new chimney, which was joined to the existing chimney with an attractive arched structure. The engines were built by Worthington Simpson of Newark and were named *Sir William Prescott* and *Bessie Prescott* after the Chairman of the Board and his wife. The pumping station was opened in October 1929 by Arthur Greenwood, the Minister of Health. These impressive engines were shut down in the 1960s and a Trust has been formed with a body of volunteers working to restore one of the engines to steam.

18. Water Supply to Richmond and Hampton Court Palaces

HEW 1437

The remains of two early water supply systems exist in south-west London and both relate to Royal Palaces on the Thames. The earliest is the *White Conduit* (TQ 190 738), which is one of three conduit houses built to serve Richmond Palace after the fire of December 1499 and dates from *c.* 1500. The *Red Conduit* and *Petersham Common Conduit* are now buried.

Just to the south of these are three brick conduit houses, dating from the 1520s, and built to serve Hampton Court Palace. They are *Combe Springs* (TQ 205 699), *Gallows Conduit* (TQ 200 702) and the *Ivy Conduit* (TQ 201 701), which is Listed Grade II and is a Scheduled Ancient Monument.

Freshwater springs on Coombe Hill, Surrey, were tapped and collected by brick-built feeders leading to the three conduit houses. From these two 3 in. diameter lead pipes conveyed the water to Hampton Court Palace almost 3 miles away. An interesting feat of construction was the laying of pipes across the River Thames. References to lead pipes and 'cokkes' for 'the counndithe' occur in the Hampton Court building accounts in 1530 and 1531. Surveys, made near Hampton Court in the 1850s for the Office of Works by 'G. H. Andrews, Civil Engineer' survive in the Public Record Office (PRO WORK 34/1330).

COLVIN, H. *History of the King's Works*, vol. 4. HMSO, London, 140.

19. Kew Bridge Pumping Station

HEW 278
TQ 187 781

These works, in *Green Dragon Lane*, Brentford, were built by the Grand Junction Waterworks Company, which was formed in 1811. The Company is of interest as it is an unusual example of a canal company becoming involved

Kew Bridge Pumping Station

KEW BRIDGE STEAM MUSEUM

Kew Bridge Pumping Station, Cornish engine house

DENIS SMITH

in public water supply. The Company's original works were in Paddington and drew water from their own canal. The Company began to draw water from the Thames at Chelsea in 1820, but removed their intake works to Brentford in 1838. The water drawn at Kew Bridge was pumped to reservoirs at Paddington, but these were abandoned when the reservoir on Campden Hill was opened in 1845. After the 1852 Water Act the Company was obliged to remove their intake above Teddington Lock and new works were opened at Hampton, Kew

Kew Bridge Pumping Station, 1837 Engine House roof

MALCOLM TUCKER

Bridge being retained as a filtration and pumping station. The works ceased operating in 1944.

Of the surviving buildings three are of particular interest: the 1837 Engine House, the 1867 Standpipe Tower, and the building housing the 90 in. and 100 in. Cornish engines. The 1837 two-storey engine house is of stock brick with a stone portico. It has an early type of iron roof, bolted together from wrought-iron flat bars, the 3 in. deep rafters being stiffened laterally by the iron roofing battens to which they are tightly wedged. The boiler house roof, spanning 23 ft on cast-iron beams, has the rafter joined at the apex by a cast-iron junction box, a detail patented in 1815, and they are trussed in a W-configuration. The other roof is a quite complex framework of narrow bars, ultimately derived from a queen-post truss. This building houses the 64 in. Boulton & Watt West Cornish engine of 1820, the 65 in. Maudslay beam engine of 1838, and the 70 in. 'Bull' engine of 1859 built by Harvey of Hayle. The 190 ft tall Standpipe Tower is of stock brick with recessed panels and stone arches and cornice. Two iron standpipes are within the brick enclosure. The steam engines pumped to these standpipes against the head of water, which then flowed by gravity onto the district. The large Cornish engine house is of rendered brickwork with round-headed windows set within recessed panels and contains the two large Cornish Engines. The 90 in. engine was built by Sandys, Carne & Vivian in 1846 and is still steamed. Alongside is the 100 in. engine built by Harvey of Hayle in 1871; this engine awaits restoration to steaming condition. The museum has also acquired waterworks machinery from a number of Greater London sites.

The buildings and machinery are now in the care of The Kew Bridge Engines Trust (formed in 1973) and the museum, with its unique collection, is open every day, and in steam every Saturday, Sunday and Bank Holiday (except Christmas and Good Friday).

20. Honor Oak Reservoir

HEW 2219
TQ 354 746

The Kent Waterworks Company, formed in 1809, was unfortunate in having the largest land area with relatively few, widespread, customers. This magnificent covered service reservoir, lying under a golf course in Honor Oak, is constructed with a rectangular grid of brick piers and vaulted ceilings. It was built under an Act (57 & 58 Vict. c.163) of 1894 obtained by the Kent Waterworks Company, and another of 1906 gave powers to the Metropolitan Water Board to complete the work.

The reservoir is 824 ft long, has a maximum width of 587 ft and occupies an area of 14.5 acres, including banks. The reservoir was at that time the largest covered reservoir in the world constructed at one time under one contract. It is divided into four sections, each of which can be filled or emptied independently. The bottom of the reservoir comprises inverted arches in concrete. The outer retaining walls are also of concrete with an inside lining of brickwork. Its total capacity is about 56.5 million gallons. Over 16 million bricks were made on site from the 173 000 cu. yd excavation. An average of 400 men worked on site over a period of three years.

The Engineer was Sir James Restler and the contractors were J. Moran & Son. The cost, including land, was £236 000 and the works were inaugurated in May 1909 by the Lord Mayor of London. The underground reservoir was named the 'Beachcroft' reservoir after Sir Melvill Beachcroft, the first Chairman of the Metropolitan Water Board, and is still in use by Thames Water.

21. Thames Water Tunnel Ring Main

HEW 2284

The aim of the late 1980s concept of Thames Water's London Water Ring Main was to improve the distribution of water throughout London and to achieve large economies by the reduction of back-pumping. It was a massive civil engineering project costing some £200 million

Thames Water Tunnel Ring Main

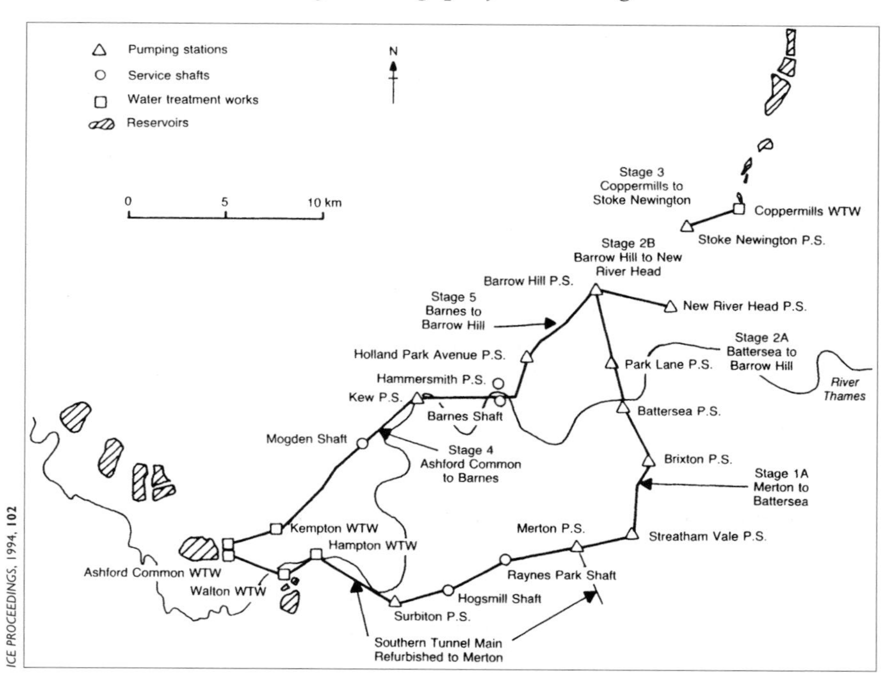

ICE PROCEEDINGS, 1994, **102**

pounds. Sixteen vertical shafts were sunk and the Water Ring Main tunnel connects these shafts at an average depth of 40 m—safely below the London Underground system. The shafts to the north of the Thames are at Ashford Common, Kempton, Kew Bridge, Barrow Hill, Park Lane, New River Head, Stoke Newington and Coppermills in Walthamstow. Those to the south are at Walton, Hampton, Surbiton, Merton, Streatham and Brixton. The tunnel extends to 80 km and is 2.54 m in diameter, and is formed of cast concrete segments. The first section of tunnel, from Ashford Common to Barrow Hill, was completed in January 1991. The Water Ring Main was designed to deliver 285 million gallons per day to a population of 5.5 million. The whole system is managed from a central control centre based at Hampton, with computer control monitoring reservoir levels pressures and remote controlled valves.

The Southern Main Drainage System

22. Deptford Pumping Station

HEW 2237
TQ 375 769

This lift station, on the Southern Main Drainage, was the first to be completed in May 1864. The original plant comprised four beam engines totalling 500 hp and ten Cornish boilers by Slaughter, Gruning & Co. of Bristol. The pumps lifted 123 million gallons per day through 18 ft. It is a stock brick building, with round-headed windows, a slate roof and a square brick chimney. Coal was delivered from Deptford Creek and stored in a cast-iron arcaded, covered coal store, which still exists. The building contractors were Aird and Son. The steam engines were replaced by reciprocating oil engines and subsequently by electric motors. It is still an operational station maintained by Thames Water.

23. Southern Outfall and Crossness Pumping Station

HEW 1115

The Southern Outfall (TQ 375 769 to TQ 483 805) is a circular brick sewer, 11 ft 6 in. diameter, which runs from Deptford pumping station to Crossness, with a fall of 2 ft per mile over a distance of 7.5 miles. To maintain communication between Deptford and Crossness an electric telegraph wire was installed in the invert of the sewer. The outfall discharged into a 6½ acre covered reservoir on the Crossness site and discharged into the Thames at suitable states of the tide. The Outfall was completed in June 1862. Sir Joseph Bazalgette was Engineer, John Grant Resident Engineer, and the contractor was William Webster.

Crossness Pumping Station (TQ 485 811) is the terminus of the Southern Main Drainage and is situated at the end of *Belvedere Road* in Thamesmead. Construction work on site began in the autumn of 1862. William Webster was the contractor for the Engine House and the covered reservoir. The river wall and large foundation works required extensive piling, the excavation of 160 000 cu. yd, and the placing of 82 000 cu. yd of mass concrete. The impressive Engine House contains four large beam engines by James Watt & Co., and they were named *Victoria, Prince Consort, Albert Edward* and *Alexandra.* The engines each originally had a single cylinder 4 ft diameter with a 9 ft stroke. The engines each have a 40 ft cast-iron beam and a 27 ft diameter fly-wheel weighing 50 tons. Unusually, the engines were subsequently compounded

Crossness Pumping Station

CROSSNESS ENGINES TRUST

DENIS SMITH

Crossness Pumping Station, Prince Consort Engine under restoration

and later even converted to triple-expansion working. In the central light-well superb decorative cast-iron screens include the monogram 'MBW' and enclose the opening to the spiral staircase leading down to the pumps at the suction level. The staff included a Superintendent and 51 others. The remoteness of this site made it necessary to have extensive housing on site, with a schoolmaster and sewing mistress on the payroll. The pumping station was opened by HRH the Prince of Wales on 4 April 1865, when an enormous company was entertained to lunch in the then unused covered settling reservoir. The Crossness Engines Trust and a group of volunteers are working to restore one of the engines to steam.

The Northern Main Drainage System

24. Abbey Mills Pumping Station

HEW 1114
TQ 387 832

This elegant pumping station is sited alongside the Northern Outfall in *Abbey Lane*, Stratford. It is cruciform in plan and is surmounted by a central lantern. In July 1865, William Webster was given the contract for the foundations, work began on site, and it was not until April 1866 that the Metropolitan Board approved the architect's drawings. In order to save time Webster was asked to continue with the structure on a schedule of prices basis, after he claimed to be able to complete the building and roof within 12 months. The station housed eight beam engines, two in each arm, and the contract for the engines, 16 boilers and the elaborate cast-iron structural work was let to Rothwell & Co. of Bolton. The original staff comprised 26 men. The beam engines were removed in 1936 and were replaced vertical-spindle centrifugal

Abbey Mills Pumping Station

GREATER LONDON COUNCIL

pumps, which are still used occasionally. The two free-standing chimneys were taken down at the beginning of the Second World War. Thames Water now retain the 1865 building for storm water pumping only and have replaced it with a new station, opened in 1997, on an adjacent site. The new pumping station has a reinforced-concrete substructure containing the low-level culverts, with the pumps on the ground floor. The roof comprises bowstring rafters and the whole building is clad in gleaming metal. The structural engineers were Ove Arup & Partners and the architects Allies & Morrison.

25. The Northern Outfall and Beckton Works

HEW 43

The Northern Outfall Sewer (TQ 370 840 to TQ 450 822) is unlike its southern counterpart in that it is contained within an earth embankment, some 12 ft high. It runs from Old Ford, Abbey Mills Pumping Station, and to the Beckton Sewage Treatment Works (TQ 447 823) on the river Roding at Barking Creek. The outfall is constructed of brick barrels, carried in an earth embankment, and where the outfall crosses roads, canals, rivers or railways the sewer is constructed of metal pipes, originally cast iron and now steel, and supported by steel-plate girder bridges with the original decorative cast-iron balustrades. A road runs along the top of the outfall embankment which Bazalgette considered would be a useful means of communication between the pumping station and the treatment works at Beckton and also 'a road of much importance to the extending districts at the eastern extremity of the metropolis'. It is still accessible to pedestrians and cyclists. Beckton Works is the terminus of the Northern Main Drainage and originally merely impounded the flow which was released into the Thames at the beginning of the ebb tide. The London County Council undertook major works here in the 1950s, and by 1955 six detritus channels, a screen house and primary sedimentation tanks had been built. From 1955 to 1959 a diffused air activated sludge plant was constructed to deal with 60 million gallons per day, together with a power house containing eight turbines running on sludge gas or oil. The £10 million works extension was opened by the Duke of Edinburgh on 22 October 1959.

26. The Western Pumping Station

HEW 2238
TQ 287 780

In *Grosvenor Road*, Pimlico, this lift-station is at the head of the Northern Low Level sewer and was designed to drain the 'Western District' area of 14½ square miles. Work on site began on 26 July 1873 and it was opened on 5 August 1875. The pumping station is situated at the end

Western Pumping Station

DENIS SMITH

of the Grosvenor Canal and on the site of the old Chelsea Waterworks. Its architecture resembles a French chateau and is roofed with solid copper tiles. The panelled square stock-brick chimney, 172 ft tall, is still a west London landmark. The contract price for the buildings was £126 955 and that for the pumping machinery was £56 789. The original plant comprised four single-acting beam engines, with riveted wrought-iron beams, by James Watt & Co., each of 90 hp. Plunger pumps (two per engine) lifted the dry weather flow 18 ft into the sewer. Steam at 40 lbf/sq. in. was produced in eight boilers. The beam engines ceased working at 2.45 p.m. on 24 April 1935 and were demolished after very nearly 60 years' work. The steam plant was replaced, without modifying the building, by oil engines supplied by W. H. Allen of Bedford, and they remain in the building today. The station is still operational using electric pumps. The opening of this station in 1875 marked the completion of Bazalgette's system of Main Drainage.

PETREE, J. F. The Western Pumping Station, Pimlico. *The Engineer*, 1935, **159**, Pt. 1, 679.

27. Markfield Road Pumping Station, Tottenham

HEW 2239
TQ 344 889

This is a good example of a non-Metropolitan Board of Works drainage scheme. The Tottenham & Wood Green Joint Drainage Board built this sewage pumping station in 1886. The handsome compound beam engine is by Wood Brothers of Sowerby Bridge, Yorkshire. It is of the self-contained type, with superior architectural and sculptural qualities, having cast-iron fluted columns, elaborate pedestals, entablature and brackets, and a foliate pedestal supporting the centrifugal governor. Unusually, the 21 ft long beam is fabricated from riveted wrought-iron plates. The flywheel is 27 ft in diameter. The 100 hp beam engine drove two plunger pumps each 2 ft in diameter and of 4 ft 3 in. stroke and at 16 rev./min delivered 4 million gallons per day via the London County Council system in Hackney and the Northern Outfall sewer to the works at Beckton. The pumping station was closed in February 1964, and is now in the

DENIS SMITH

Markfield Road Pumping Station, Tottenham

care of a trust body that has steamed the engine on open days in recent years. The engineer to the Joint Drainage Board was Colonel P. E. Murphy, MICE, and the Borough Engineer and Surveyor of Tottenham, W. H. Prescott, acted as consulting engineer.

PRESCOTT, W. H. The local administration of Tottenham and its development. *Inc. Assoc. Mun. County Engrs*, 1905–06, **32**, 165–166.

28. West Ham Pumping Station

HEW 1116
TQ 389 832

This sewage pumping station was built under powers granted by the West Ham Corporation Act of 1893 and is bounded by *Abbey Road* to the north, *Canning Road* to the east, the Channelsea River to the west and the Northern Outfall Sewer to the south. The steam-powered plant comprised two beam engines, by the Lilleshall Company of Shropshire, three direct-acting engines driving centrifugal pumps, four hand-fired Lancashire boilers and five mechanically stoked Lancashire boilers. The compound beam engines have cylinders of 2 ft 6 in. and 4 ft diameter. Each engine has a cast-iron beam 30 ft long and weighing 17 tons and a flywheel of 22 ft diameter. The total capacity of the beam engines was 2 million

DENIS SMITH

West Ham Pumping Station

gallons per hour, and that of the centrifugal pumps 4 826 000 gallons per hour. The beam engines worked alternately dealing with dry weather flow and a certain amount of rainfall and they delivered into the adjacent barrels of the Northern Outfall, then owned by the London County Council. The works was opened in 1900 and was closed in the 1970s, but the two beam engines remain in the building. The building is Listed Grade II, although the future of the beam engines is uncertain.

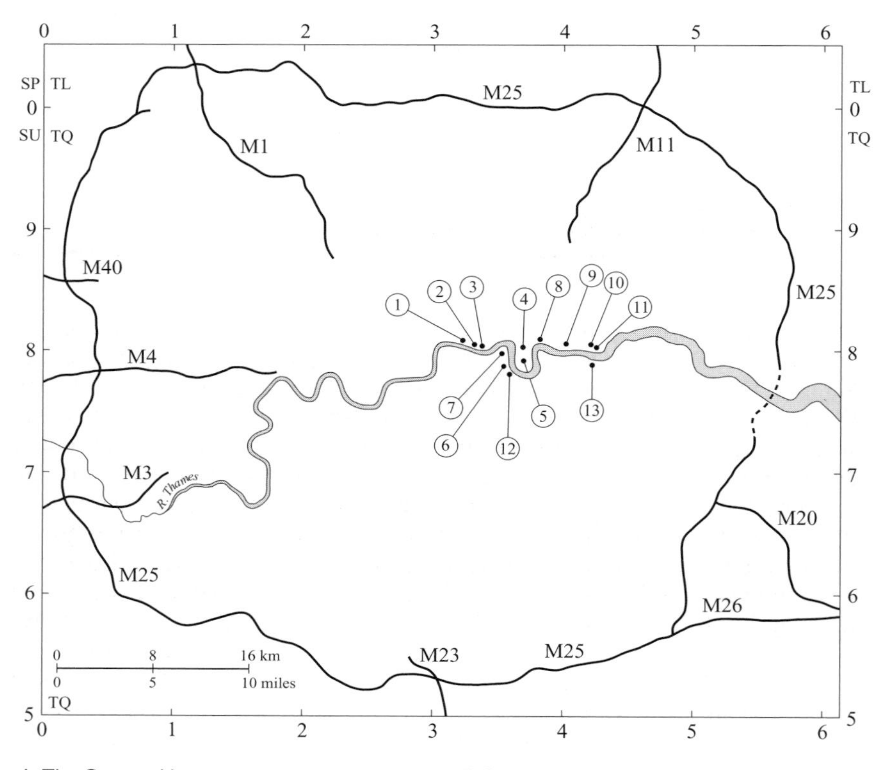

1. The Custom House
2. St. Katharine Docks
3. London Docks and the New Tobacco Warehouse
4. West India Docks
5. Millwall Docks
6. Howland Great Wet Dock
7. Surrey Commercial Docks
8. East India Docks
9. Royal Victoria Dock
10. Royal Albert Dock
11. King George V Dock
12. Deptford Royal Naval Dockyard
13. Woolwich Royal Naval Dockyard

3. The Port of London

The Thames has always been of strategic importance, as a commercial artery, for its Naval Dockyards, and as a town planning boundary. Since Roman times the Thames had been used for the transport of goods. Traditionally, up to the end of the eighteenth century, London's river trade was handled by large numbers of relatively small sailing vessels. These craft would moor in the river, between Old London Bridge and the Tower ('The Pool of London') and discharge their cargo, over-side, into lighters and the lightermen would then row ashore to offload their goods onto riverside wharves. Towards the end of the eighteenth century congestion in the Pool became intolerable. From 1700 to 1770 the trade of the port doubled, and from 1770 to 1795 it doubled yet again. In 1796 the Government set up a Parliamentary Committee appointed 'to enquire into the best mode of providing sufficient accommodation for the increased trade and shipping of the port'. In 1799 Parliament authorised the construction of a dock on the Isle of Dogs and the preamble to the Act says:

> ...the ships in the West India Trade frequently arrive at the Port of London in large fleets, and occasion great crowding, confusion and damage therein; that great obstruction and delays arise from their cargoes being carried in lighters to the legal quays, and that in the passage thither, such cargoes are subjected to pilfering and fraud, whereby the owners sustain great loss, and the revenue is much injured...[1]

This dock became known as 'The West India Dock' and led to a flurry of activity in the planning of a system of riverside wet docks in the early nineteenth century. This massive programme of civil engineering involved many of the great civil engineers of the period, including Ralph Walker, William Jessop, John Rennie, Thomas Telford, James Walker, George Parker Bidder and many others. Once the system was established civil engineers were involved during the nineteenth century in responding to developments in naval architecture, shipbuilding and power technology. The change from wooden to metal hulls, from sail to steam and from paddle-wheel to screw propulsion all meant that vessels became larger and could carry more cargo. Locks between the Thames and the wet docks had to be widened and deepened and warehouse

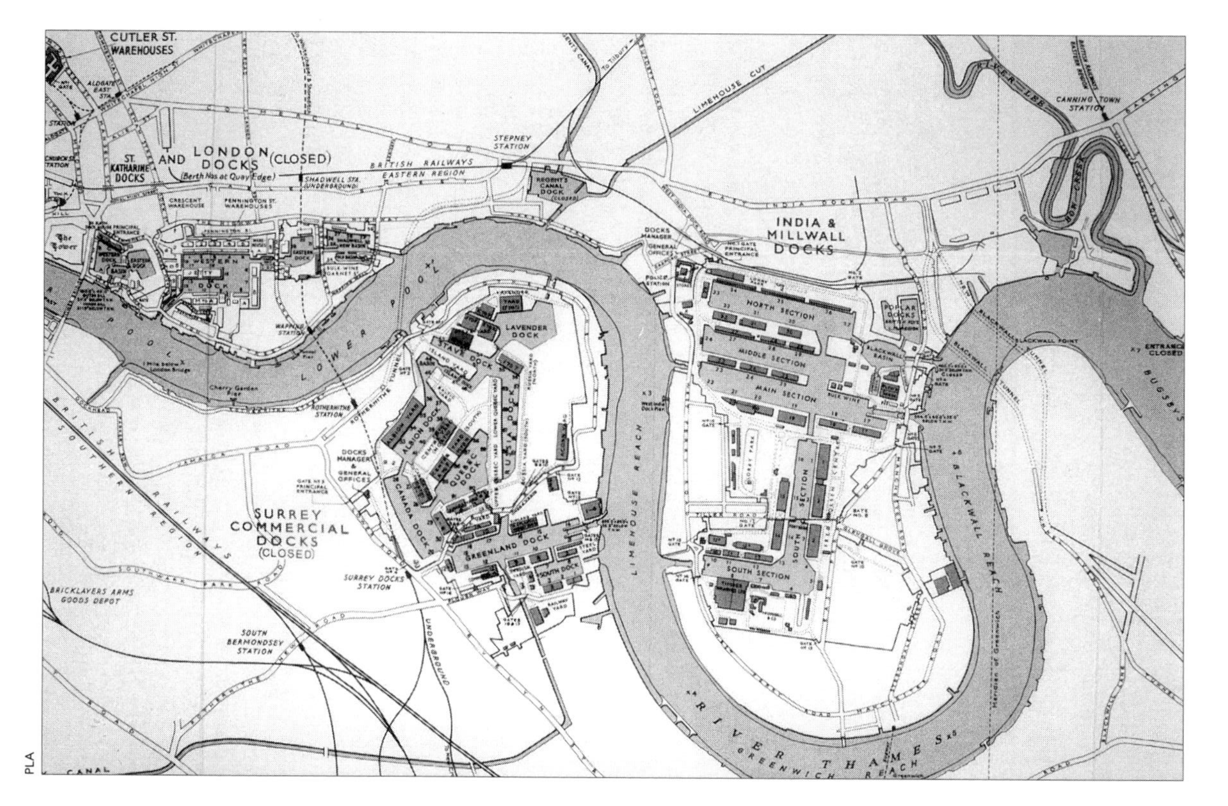

PLA

Plan of St. Katharine, London, Surrey Commercial, India and Millwall Docks

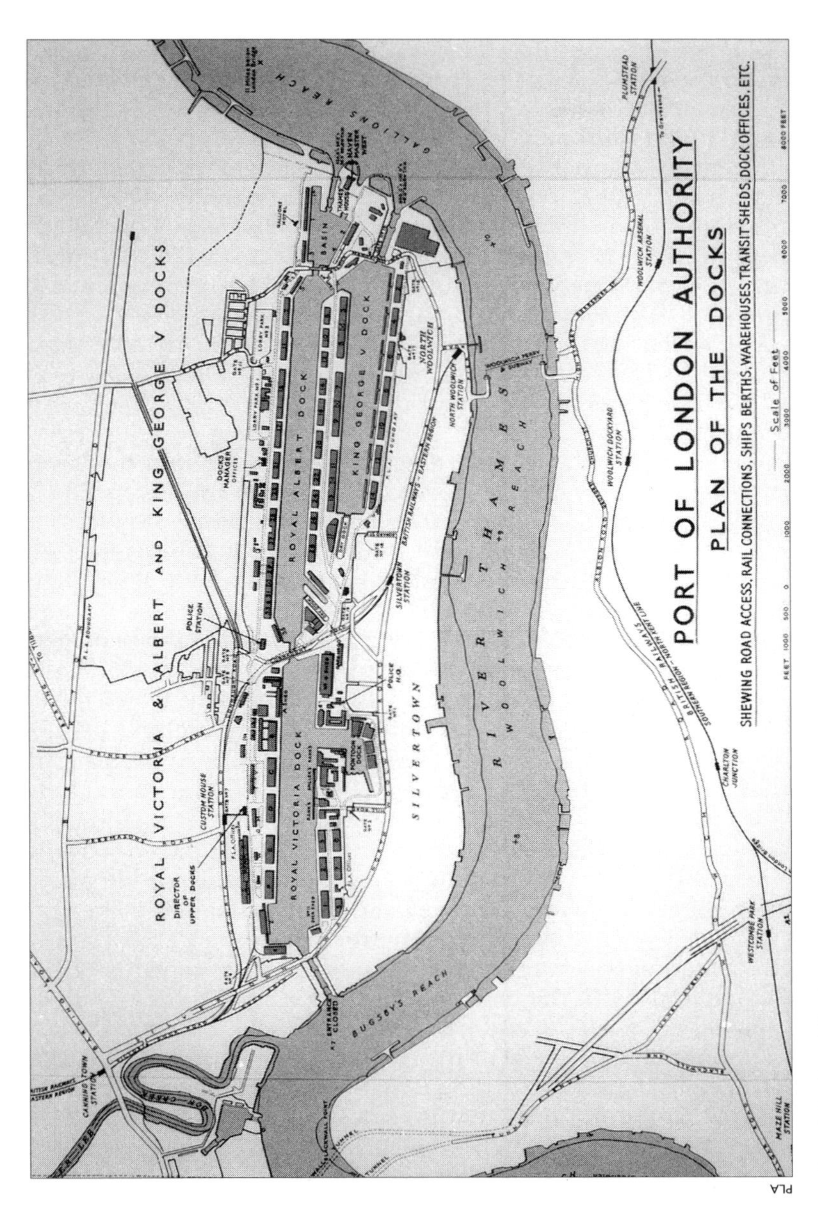

PLA

Plan of the Royal Docks

accommodation had to be greatly increased, all leading to major civil engineering works. With the advent of hydraulic power from the middle of the nineteenth century the dock companies were quick to install their own systems supplying pressurised water from pumping stations through mains to operate warehouse cranes, capstans, lock gates, and for opening and closing moveable bridges. Towards the end of the nineteenth century another technological development made its impact on the port, namely the import of refrigerated meat, although there had long been a large trade in imported ice. In 1875 a small quantity of chilled beef (packed in ice) was shipped, from New York to Britain, and in 1879 a refrigeration machine was first fitted in a ship. The experimental stage began in 1881 and was becoming fully established by 1886–87. The London & St. Katharine Dock Company was amongst the first to respond to the frozen meat trade by constructing refrigerated storage facilities at the Royal Victoria Dock in 1881, and the East and West India Company quickly followed.

The management of the port involved many different bodies. First, there were the private dock companies, and by 1809 there were nine such companies, including (with the date of incorporation) the West India (1799), the London (1800), the East India (1803), the Commercial (1807) and the Baltic (1809). The first amalgamation occurred in 1810 when the Baltic Dock Company was absorbed by the Commercial Dock Company. By the end of the nineteenth century major reform of the management structure of the port was being discussed. A Royal Commission report in 1902 led to the formation The Port of London Authority, which took control in 1909.Their jurisdiction extended from Teddington Lock to the Thames Estuary on a line joining Havengore Creek on Foulness island to Warden Point on the Isle of Sheppey. The duties of the Port of London Authority included all matters relating to navigation, regulation of river traffic, maintenance of river channels, provision and upkeep of a number of public ship and barge moorings and the licensing of wharves and structures that extend into the river below high water mark. The Authority's income was derived principally from dues charged on ships and cargoes entering the river and docks and charges to merchants for services connected with the handling and storage of cargoes. The Port of London has its own local datum for levels known as 'Trinity High Water'. This was the mean height of ordinary Spring Tides and was 11.4 ft above the Ordnance Datum at Newlyn. Other bodies involved in running the port were HM Custom and Excise levying duties on goods and the Trinity House river pilots who were responsible for navigating ships upstream from Gravesend into the port. Gravesend is also the base for HM Customs Waterguard and Preventive Officers and the station of the Port Medical Officer of Health. Customs and Excise also affected civil engineering

Table 3: The five dock systems of the Port of London

Dock system	Land area (acres)	Water area (acres)
London and St. Katharine	125	45
Surrey Commercial	381	136
India and Millwall	515	155
The Royals: Victoria, Albert, King George V	1112	235
Tilbury	725	104
Total	2858	675

design by stipulating in the 1790s separate import and export docks at the West and East India Docks. Developments at Tilbury, and in cargo handling generally, led to the closure of many of the upstream docks in the late 1960s. The East India Dock closed in 1967, St. Katharine Docks in 1968, London Dock in 1969 and the Surrey Commercial Docks in 1970.

The Port of London under the Port of London Authority was ultimately divided into five separate dock systems; an overview of the systems and their relative areas is given in Table 3.

The London Docklands Development Corporation was inaugurated in 1981, and with the re-development of Docklands in recent years much has been demolished. Nevertheless, surviving examples include examples of dock engineering, locks and lock gates, warehouses, moveable bridges and cranes, many of which involved hydraulic power systems. In addition to work in the commercial port, civil engineers were often employed in the development and maintenance of the Royal Naval Dockyards at Deptford and Woolwich on the Thames. The Historical Engineering Works in the port are here described geographically from west to east.

[1]39 Geo. 3 c.69 (1799).

WILLIAMSON, E. and PEVSNER, N. *London Docklands; An Architectural Guide*. Penguin Books, Harmondsworth, The Buildings of England series, 1998.

CARR, R. J. M. (ed.). *Dockland: An Illustrated Historical Survey of Life and Work in East London*. NELP/GLC, London, 1986.

GREEVES, I. S. *London Docks 1800–1980: A Civil Engineering History*. Thomas Telford, London, 1980.

SKEMPTON, A. W. Engineering in the Port of London, 1808–1834. *Trans. Newcomen Soc.*, 1981–82, **53**, 73–96.

SKEMPTON, A. W. Engineering in the Port of London, 1789–1808. *Trans. Newcomen Soc.*, 1978–79, **50**, 87–108.

BIRD, J. *The Major Seaports of the United Kingdom*. Hutchinson, London, 1963.

1. The Custom House

HEW 2326
TQ 332 806

The Custom House has its north elevation in *Lower Thames Street* with its south elevation overlooking the Pool of London. With its associated quay it has been crucial to the maritime commerce of the Thames since at least the fourteenth century. The building of 1559 was destroyed in the Great Fire of London in 1666. Sir Christopher Wren designed a large replacement Custom House, built in 1669–71, in which the offices were grouped around a 'Long Room'—a name which survives to the present day, both in London and in Custom Houses in the previous British Colonies. Wren's building did not survive long as it was irreparably damaged by fire in 1715. The subsequent Custom House was built, on Wren's foundations, to the designs of Thomas Ripley in 1718–25. Trade in the port of London grew rapidly in the late eighteenth century, and in 1809 the Customs service sought parliamentary powers to build a new building to the west of Ripley's structure.

The present building has its origins in a design by David Laing, who was appointed Surveyor of the Customs in 1810. Treasury approval was granted in 1812 for Laing's building, estimated to cost £228 000. The contract was let to John Mills and Henry Peto and work began in October 1813. However, there were many delays and the building was not completed until late 1817. Jolliffe & Banks built the granite quay and river stairs at the same time to the designs of John Rennie. During the early 1820s subsidence had caused serious damage, culminating in the collapse of the Long Room floor in January 1825. Sir Robert Smirke was retained to deal with the problems and he replaced nearly all the central block with structurally innovative work, including concrete foundations and both structural and fire-resistant ironwork. Smirke installed cast-iron columns and beams of up to 30 ft span and over his 190 ft by 63 ft Long Room he use arched wrought-iron fireplates over the ceiling to increase fire resistance. The ironwork was supplied by Foster, Rastrick and Company of Stourbridge.

The Custom House underwent extensive refurbishment in 1992, with Hurst, Pierce and Malcolm as the structural engineers, who commissioned both documentary research and measured drawings of the structural elements of the historic building, all held by the Royal Commission on the Historical Monuments of England.

ROYAL COMMISSION ON THE HISTORICAL MONUMENTS OF ENGLAND. *The London Custom House*. Royal Commission on the Historical Monuments of England, Swindon, 1993.

2. St. Katharine Docks

HEW 58
TQ 349 805

Phillip Hardwick was architect for the warehouses and the St. Katharine Dock House. The project involved the displacement of some 1200 people and the demolition of St. Katharine's church, and a massive excavation. The docks were constructed in 1826–29. This scheme, adjacent to the Tower of London, was designed by Thomas Telford to provide a secure dock, which he said was designed to reduce 'Lighterage and Pilferage'. The contractors, Bennett and Hunt, began excavation in May 1826 and completed the work in 18 months. The total construction took just two and a half years in spite of the failure of the river cofferdam in 1827. Captain Carlsund of the Swedish Navy visited the site and described the work thus:

> I frequently witnessed a thousand men and several hundred of horses employed in the operations .... At the beginning of the works wheelbarrows were employed to carry away the earth, but as excavations proceeded and became deeper, iron railways and steam-engines were substituted.

St. Katharine Docks

WENDIE TEPPETT

The excavated soil was taken by barge to the Duke of Westminster's estate in Pimlico. The scheme comprised an *Entrance Lock*, 180 ft long, 45 ft wide and 25 ft deep, an *Entrance Basin* and the *East* and *West Docks*. Warehouses were built directly alongside the dock walls to facilitate the direct transfer of goods from ship to warehouse. In 1957 the entrance lock was rebuilt, with new gates. The *Dockmaster's House* survives beside the entrance lock and the *Ivory House*, built in 1858–60, with its clock turret sits between the two docks.

An interesting survival is the small footbridge on display close to its original site. The port archives contain a drawing of the bridge which is dated October 1829 and signed by Thomas Rhodes, Telford's Resident Engineer at the dock from November 1828. On 27 October 1829 the dock company accepted the tender of £446 from John Lloyd, a Westminster millwright, for building a wrought-iron footbridge between the Basin and the Eastern Dock. For access to the dock the two halves of the bridge were withdrawn into recesses in the masonry abutments. Schemes to strengthen the bridge were visually intrusive and were rejected in favour of removal and preservation. The bridge remained in its original position until 1993. It is not only a link with the early period of St. Katharine Docks, but is also a rare survival of this type of structure.

St. Katharine Docks were sold to the Greater London Council for £1.7 million and the design competition for redevelopment of the site was won by Taylor Woodrow.

SIMMS, F. W. *The Public Works of Great Britain*. John Weale, London, 1838.

3. London Docks and the New Tobacco Warehouse

HEW 1567
TQ 347 806

The promoters of this dock consulted John Rennie, Robert Mylne and Captain Huddart in1798 and in 1800 obtained an Act to construct 'wet docks as near as may be to the City of London and seat of commerce'. The company was granted a 21-year monopoly for handling ships laden with tobacco, wine, rice and brandy (except for ships from the East and West Indies). Rennie was appointed engineer in May 1801. By January 1802 much of the site had been cleared and excavation begun. On 26 June 1802 foundation stone ceremonies were witnessed by the Prime Minister, the Chancellor of the Exchequer and 'genteel persons of both sexes'. In January 1803 the *Western Dock* (20 acres) was under construction and the cofferdam for the *Wapping Basin* (3 acres) was complete. The first ship entered the dock on 31 January 1805, although the warehouses were not complete. Daniel Asher Alexander was architect and surveyor to the company and he designed the five Pennington Street warehouses (demolished in 1979) on the north side of the *Western Dock* and also the *London Dock House*. The Wapping basin was the original entrance to the Dock, but this was infilled in 1956. The *Tobacco Dock* was built in 1811–13 by Rennie, facilitating second-phase development to the east. The *Eastern Dock* (1824–28) was designed by William Chapman as consulting engineer. *Shadwell Old Basin* (1828–32,) by J. R. Palmer, provided an eastern entrance until the *Shadwell New Basin* (1854–58) by J. M. Rendel completed the London Docks scheme. The docks were roughly infilled in the 1970s and the Western Dock again infilled in 1980 to provide a housing site. The London Docks are divided by *Wapping Lane*, *Garnet Street* and *Glamis Road*, each of which still contains its lifting bridge.

The New Tobacco Warehouse (TQ 347 806) is a fascinating structure, and the sole survivor of the warehouses on the north quay of the London Dock. It was built in 1811–14, although it has been modified in recent years for commercial purposes. It was designed by D. A. Alexander as a single-storey bonded warehouse storing tobacco on the ground floor, having an original area of 210 000 sq. ft, with wine and spirit stored in the stone-groined vaulted basement. About 60% of the floor space survives. The warehouse was later used to store hides and became known as 'The Skin Floor'. The impressive roof comprises timber queen post trusses, of 53 ft 10 in. span, with two scarf joints in each tie member. Rising from the two queen posts a central glazed clerestory runs the length of the building. Wrought-iron strap work is used to strengthen the timber joints. Alternate roof trusses are supported on cast-iron columns of cruciform section, at 18 ft centres, and in each third bay a hollow column acts as a rainwater pipe. Cast-iron raking struts from the columns support the intermediate roof trusses and at right angles Y-form struts carry the valley gutter. The vast ground floor conveys the effect of a large petrified forest. The basement has stone piers, on an 18 ft square grid supporting the cast-iron columns above, with groined vaults with a maximum headroom of 9 ft supporting the stone-flagged ground floor. The conversion of the building to the Tobacco Dock shopping centre was undertaken by the Terry Farrell Partnership in 1984–89.

4. West India Docks

HEW 2206

This was the first part of the new wet dock system to be constructed under Ralph Walker with William Jessop as Consulting Engineer. The West India Dock Company received their Act in 1799 and the system was opened 1802–06. The dock accommodation comprised the *Import Dock* (TQ 375 805) the *Export Dock* (TQ 375 802) and the *South Dock* (TQ 376 800).

The most important surviving buildings are the magnificent group of warehouses at the western end of the North Quay. The warehouses and offices were built in 1802–13 as one large block by the architect George Gwilt. *No. 2 Warehouse* (TQ 373 806) of 1802 is the first multi-storey warehouse of the London enclosed docks, and the cast-iron columns of 1814 are the earliest surviving in a multi-storey warehouse. *No. 1 Warehouse* (TQ 372 806) was built in 1803 and was extended to its present height in 1827. Blocks A to E have been converted for residential and commercial use, but the works to blocks F to H are allocated to the proposed Museum in Docklands. This warehouse complex is Listed Grade I.

5. Millwall Docks

HEW 2207

In 1864 an Act incorporated the 'Millwall Freehold Land and Canal Company', which was established to build a dock to provide waterside sites for the increasing heavy industry of the Isle of Dogs. The docks were designed by John Fowler and William Wilson and the contractors were Kelk and Aird. Work began in 1865 and the docks were opened in March 1868. In 1866 the failure of the Overend & Gurney bank led to the failure of several Thames shipyards and other industries. This caused difficulties for the Millwall Dock Company in its early years and it had to work hard to generate trade for the docks, but it soon secured a large share of the grain trade and became the leading grain dock in the country. The Millwall *Inner Dock* (TQ 377 794) and the Millwall *Outer Dock* (TQ 375 791) are virtually at right angles to each other. The water area in these docks remains virtually untouched by the redevelopment activity in the Isle of Dogs.

SARGENT, E. Frederick Eliot Duckham, MICE, and the Millwall Docks (1868–1909). *Trans. Newcomen Soc.*, 1988–89, **60**, 49–71.

6. Howland Great Wet Dock

HEW 2205
TQ 363 792

Built under an Act of 1696 this dock was completed in 1699. Named after a south London landowner, John Howland, it was a timber-walled dock entered through a wooden lock from the Thames. It was originally used as a sheltered mooring with avenues of trees forming a windbreak. The dock was designed by John Wells, a local shipwright, who also supervised its construction. William Ogbourne, a house carpenter from Stepney, was the contractor. Ships were built here and from the 1720s it served the whaling fleets with boilers and tanks to extract sperm oil from blubber. In 1763 the dock was sold and renamed the *Greenland Dock*. It was acquired by the Surrey Commercial Dock Company in 1807, when Ralph Walker built a new entrance lock. Sir John Wolfe Barry rebuilt the dock to its present size (22½ acres) in 1894–1904.

The bust of James Walker standing on its pedestal on *Brunswick Quay* (TQ 361 792) was commissioned by the Institution of Civil Engineers from the sculptor Michael Rizzello and was unveiled by the Institution's President in 1990.

SKEMPTON, SIR A. W. (ed.). *Biographical Dictionary of Civil Engineers in Great Britain and Ireland: 1500–1830*. Thomas Telford, London, 2001.

Greenland Dock entrance

DENIS SMITH

7. Surrey Commercial Docks

HEW 2208

This was the only dock complex in the Port of London to lie on the south side of the river. The Commercial Dock Company was formed in 1807 to develop the site in Rotherhithe around the Greenland Dock. The early engineering work here was done under Ralph Walker, assisted by his nephew James Walker. James was appointed Engineer to the company in September 1810 at a salary of £800 a year, and his firm remained Engineers to the company until 1862, when Walker died. The scheme

DENIS SMITH

Norway Dock entrance, swing bridge (James Walker, 1855)

developed over the years with the construction of the *Baltic Dock* (1809), *East Country Dock* (1811), *Albion Dock* (1860), *Canada Dock* (1876) and *Quebec Dock* (1926). The East Country Dock became the *South Dock* (TQ 364 790) when James Walker enlarged the old dock and built a new river lock in 1851–52. Walker designed an elegant wrought-iron footbridge over the lock in 1855, and this was relocated to the new development at *Norway Dock* in 1987.

The Docks closed in 1969 having occupied 460 acres, dealing principally in timber. *Stave*, *Lavender* and *Lady Docks* were shallow basins and used only by lighters and for floating timber. The only two small areas of water left in the docks are the *Surrey Basin* (TQ 357 800), the north-west entrance to the docks, and in the north part of *Canada Dock*, now known as *Canada Water*.

8. East India Docks

HEW 2207

These docks were built adjacent to Brunswick Wharf, a wharf constructed of cast-iron plates and wrought-iron land ties by James Walker and George Parker Bidder in 1834. The docks were designed for the East India Dock

Company by John Rennie and Ralph Walker and were built in 1803–06. The docks comprised the *Export Dock* (8 acres) and the *Import Dock* ($12\frac{1}{4}$ acres), but only the *Entrance Basin* (TQ 391 880), of 1897, survives as the sole water feature. The lock gates are by the Thames Ironworks Company and the hydraulic machinery by Sir W. G. Armstrong Whitworth & Company. The *Import Dock* on the north of the site was pumped dry in 1943 to build the Mulberry Harbours, and was filled in only after the docks closed in 1967. The *Export Dock* to the south was bombed and was used for the construction of Brunswick Wharf Power Station built in 1947–56 and demolished in 1988–89. Along Leamouth Road part of the 20 ft high *Import Dock Wall* survives.

9. Royal Victoria Dock

HEW 2209
TQ 409 806

This dock is unusual in that the promoters of the scheme were the railway contractors Peto, Brassey and Betts. The Engineer was George Parker Bidder. Construction began in 1850 and the dock was opened by Prince Albert in 1855. The prefix 'Royal' was added in 1880. The dock was $1\frac{1}{4}$ miles long with 94 acres of water. A major reconstruction of the dock took place from 1935 into the 1940s. The

Royal Victoria Dock warehouse

DENIS SMITH

whole project was a contractor's speculation and cost £1 076 664.

The remaining buildings include *Warehouse K* (TQ 405 808), a bonded store for tobacco built in 1859 by George Parker Bidder, and the only remaining section of the nineteenth-century range. The interior comprises timber floors on cast-iron cruciform columns with unusual timber-trussed roof. The building is Listed Grade II. The other surviving building of merit is *Warehouse W* (TQ 418 807), built in 1883 by Robert Carr, a narrow four-storey brick structure with timber floors and cast-iron columns. The exterior has recessed brick panels with round heads at the third-storey level, and three vertical bays with loading doors to all four floors.

BIDDER, G. P. Historical notice of Victoria Docks. *Min. Proc. Instn Civ. Engrs*, 1958–59, **18**, 483.

10. Royal Albert Dock

HEW 2210

This dock was built under an Act of 1875 obtained by the London and St. Katharine Dock Company. The dock was opened by the Duke of Connaught in 1880. The Engineer was Sir Alexander Rendel, and the contractors were Lucas & Aird. The dock is 1¼ miles long and occupies 71 acres, and at the east end the *Connaught*

The Royal Docks today

DENIS SMITH

Passage (TQ 417 806) gives access to the Royal Victoria Dock. The dock walls are mostly of Portland cement concrete. At the north-west corner of the dock there were extensive cold storage facilities (*c*. 1914–17), and the red brick cold store compressor house survives. The London & St. Katharine Dock Company provided passenger facilities, and the *Connaught Tavern* (1881) (TQ 416 808), the *Gallions Hotel* (1883) (TQ 439 807), the *Central Buffet* (TQ 428 808) and the *Dock Manager's Offices* (1883) survive. The Buffet and the Offices are Listed Grade II and the Gallions Hotel is Listed Grade II*.

11. King George V Dock

HEW 2211

With the formation of the Port of London Authority in 1909 a programme of new works was proposed by Frederick Palmer, their Engineer. Detailed design work for a new dock at North Woolwich was begun in the autumn of 1911. The First World War delayed the progress of the work and the dock was opened by King George V on 8 July 1921.

The works comprised the *Wet Dock* (TQ 424 802 to TQ 438 803), of 64 acres and 38 ft depth of water and providing a total length of quay of just over 2 miles. The *Entrance Lock* (TQ 439 802), 100 ft wide and 800 ft long, had a depth over sills at Trinity High Water of 45 ft., and the *Dry Dock* (TQ 423 804) was 100 ft wide and 750 ft long. A passage, 100 ft wide and 34 ft deep, gave access to the Royal Albert Dock. A steel lattice swing bridge, operated by hydraulic power, carries a road over this passage, and a double-leaf bascule bridge carries *Woolwich Manor Way* across the Entrance Lock.

Frederick Palmer resigned in 1913 and was followed by Sir Cyril Kirkpatrick, and the resident engineer was Asa Binns. The main contractors were S. Pearson & Son Ltd. Sir William Arrol and Co. Ltd. were contractors for the lock gates, caisson, bascule and swing bridge. The warehouses on the peninsula between this dock and the Royal Albert Dock were demolished to make way for the runway of the London City Airport.

BINNS, A. King George V Dock, London. *Min. Proc. Instn Civ. Engrs*, 1922–23, **216**, Pt. II, 373.

The Naval Dockyards

12. Deptford Royal Naval Dockyard

HEW 2213
TQ 370 782

The Naval complex occupied land between *Watergate Street* and *Dudman's Dock* and comprised the Naval Dockyard and its associated Victualling Yard to the north of the site. John Rennie designed mill machinery for the Victualling Yard in the mid-nineteenth century

Three iron roofs over the shipbuilding slips were built at Deptford. They were all built by George Baker & Son, a firm that also built iron slip roofs in Portsmouth, Pembroke and Chatham Dockyards during the 1840s. The roof to slip No. 1 has been demolished, but those of slips 2 and 3 survive in use as warehouses, although the main spans have been altered. In March 1869 the Yard was sold to a Mr. Austin for £70 000, who sold 21 acres to the Corporation of London for £94 640. The City converted it into the Foreign Cattle Market where cattle were penned, some of them to be killed in the Navy's adjacent Victualling Yard.

John Evelyn (who lived adjacent to the Dockyard) records in his diary for 19 July 1661:

> We tried our *Diving Bell* or engine in the water-dock at Deptford, in which our Curator continued half an hour under water; it was made of cast lead, let down by a strong cable.

This is a relatively early reference to the practical use of a diving bell.

SUTHERLAND, R. J. M. Shipbuilding and the long span roof. *Trans. Newcomen Soc.*, 1988–89, **60**, 107–126.

13. Woolwich Royal Naval Dockyard

HEW 2212
TQ 423 791

The Royal Naval Dockyards were at one time among the largest industrial establishments in the country, employing many eminent consulting civil engineers in their construction and large numbers of craftsmen within the works. Woolwich Dockyard was established in 1513 by Henry VIII to build and maintain wooden ships for the Navy.

The Dockyard closed in 1869. Three of the large-span iron-framed building slip roofs from the 1840s were subsequently removed to Chatham Dockyard where they

Woolwich Royal Naval Dockyard, Superintendent's House

DENIS SMITH

are now known as *No. 8 Machine Shop* (Woolwich No. 6 slip), the *Boiler Shop* (Woolwich No. 4 slip) and the *Prom EW Factory* (Woolwich No. 5 slip). The oldest surviving building is the *Superintendent's House and Office*, built in 1778–84. Several nineteenth-century buildings survive, including the *Mould Loft* of 1815 and two *Dry Docks* of 1843, which are now filled with water and used for leisure pursuits. A major landmark is the octagonal stock brick chimney, 200 ft high, which is part of the *Steam Factory* of 1844, now in industrial use.

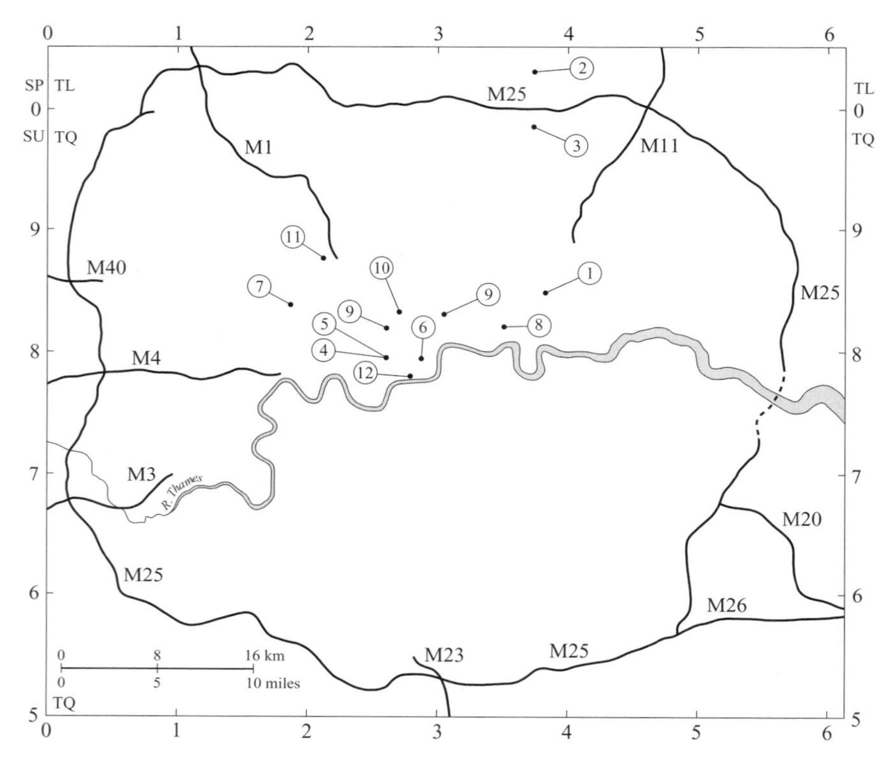

1. River Lea and Lea Navigation
2. Waltham Abbey Powdermill
3. Royal Small Arms Factory, Enfield
4. River Westbourne and the Serpentine Lake
5. The Serpentine Bridge
6. St. James's Park Footbridge
7. North Circular Road Aqueduct
8. The Regent's Canal
9. Maida Hill and Islington Tunnels
10. Macclesfield Bridge, Regent's Park
11. The Brent and Ruislip Reservoirs
12. Grosvenor Canal, Pimlico

4. Rivers and Canals

Rivers

The tributary rivers of the Thames in the Greater London area have all, at one time or another, been used for industrial or leisure purposes. Although rivers in their natural state cannot properly be regarded as part of London's civil engineering heritage, nevertheless millwrights and, later, civil engineers, have controlled the regime of these streams to harness the energy of water for mills, to improve them for transport, to extract water for drinking purposes and to increase their natural drainage characteristics and prevent flooding. On the north side of the Thames valley the rivers naturally flow south into the Thames. The principal examples of these rivers that are of engineering heritage interest are the rivers *Lea* (or Lee), *Roding*, *Fleet*, *Westbourne*, *Counters Creek* and the *Brent*. Similarly, on the south side the important rivers are the *Wandle* and the *Ravensbourne*. During the eighteenth and nineteenth centuries, as urban development increased, some of these rivers were buried in underground culverts and became known as the 'lost' rivers of London. These culverted rivers have caused problems for subsequent engineering schemes such as the construction of the main drainage system and the underground railway in the 1860s, as well as for twentieth-century and later civil engineers.

The course of many of these rivers has determined several London Borough boundaries. An interesting survival from the sixteenth century is the *Duke of Northumberland's River* (TQ 150 734 to TQ 167 759). This is not a natural river but a man-made cut, just over 2½ miles long, from the River Crane at *Kneller Gardens* in Twickenham to *Church Street*, Isleworth, where it joins the Thames. The river was undertaken at the expense of Henry VII to serve the Abbey at Syon and its mills at Twickenham and Isleworth.

The River Brent in west London flows southwards towards the Thames at Brentford. The river joined the Grand Junction Canal at Hanwell when the canal was opened in 1801. The valley of the Brent was the first major obstacle facing I. K. Brunel on his Great Western Railway route to the west from Paddington. Further north the Brent, and its tributary *Silk Stream*, was dammed to form a summit-level feeder reservoir for the Regent's Canal Company.

The River Roding in east London flows from Essex to the Thames at Barking Creek. During the late eighteenth and early to mid-nineteenth century Barking was an important fishing port serving London, and part of the Town Quay, basin and a granary building survives. The river was navigable from the Thames up to Ilford, serving riverside industries, until the late 1950s.

South of the Thames two rivers have been of prime importance to industry, namely the River Ravensbourne to the east and the River Wandle to the west. The Wandle rises in the high ground in the Croydon area, and flows south into the Thames at Wandsworth. This modestly sized river was at one time one of the most heavily used rivers in the country for industrial purposes and remains of water-powered mills survive along its banks. The Wandle valley was chosen by William Jessop as the route for the *Surrey Iron Railway*, opened from Wandsworth to Croydon in 1803 and from Croydon to Merstham in 1805. This horse-drawn tramroad reflects the importance of the extensive industry in the Wandle valley. The River Ravensborne flows north, entering the Thames at Deptford Creek. Again, the river flow was harnessed for industrial purposes. The river supplied many corn mills, The Royal Armoury and Small Arms factories at Lewisham, and the Kent Waterworks Company, of which John Smeaton was a partner in the eighteenth century, built a large pumping station at Brooksmill.

Canals

Canals, and canalised rivers, in Greater London have played an important part in London's industrial development. The River Lea, and its subsequent navigation cut, has provided access to London from Hertfordshire for agricultural and other produce for several centuries. In September 1574, Commissioners were appointed to improve the navigation of the Lea and their most important work was the construction of the timber-sided lock at Waltham Abbey (TL 383 008). This lock was the first pound lock in England to be fitted with mitre gates at both ends.

Before the nineteenth century London had only distant connections with the national developments of the canal mania period. From the Thames at Reading access to the south-west was gained through the Kennet and Avon Canal. It was the Grand Junction Canal, completed between Braunston and Brentford in 1801, which connected London with the Midlands and the national canal network. The Grand Junction Company built an arm of their canal to a terminal basin at Paddington, and in 1812 the Regent's Canal Company was formed to provide a link from Paddington to the east of London at Limehouse. Two other important London canals, to the south of the Thames, were the Croydon Canal and the Grand Surrey Canal, running south from the Commercial Docks.

FAIRCLOUGH, K. The Waltham Pound Lock. *History of Technology*, 1979, fourth annual volume, 31–44.

BOYES, J. R. R. *Canals of Eastern England*. David & Charles, Newton Abbot, 1977.

BARTON, N. *The Lost Rivers of London*. Phoenix House and Leicester University Press, Leicester, 1962, revised edition 1992, reprinted by BCA, 1993.

PRIESTLEY, J. *Navigable Rivers and Canals*. 1831, Reprinted by David & Charles, Newton Abbot, 1969.

Rivers North of the Thames

I. River Lea and Lea Navigation

HEW 2240

The River Lea has its origins in Leagrave marshes in Bedfordshire, and flows generally south joining the Thames at Bow Creek. The river was used as the boundary between the counties of Middlesex and Essex. The first Act of Parliament relating to the River Lea is dated 1425. The Lea has provided water power for many mills, including two major Government sites, the Waltham Abbey

Lea Navigation, Pickett's Lock gate capstan

DENIS SMITH

Powdermill and the Royal Small Arms Factory at Enfield. The lower reaches are tidal up to Bow Locks. Under the Lea Navigation Act of 1739 the river was brought, for the first time, under some form of unified management throughout its whole navigable length, and in 1765 the Trustees employed John Smeaton to make new cuts, increase the number of locks and create a continuous towpath. The Lea Navigation begins in the town of Hertford, at 11 ft 3 in. above sea level, and is about 26 miles long to its fall into the Thames. The Lea Navigation was promoted largely to convey grain malted in Hertfordshire to the London breweries. To shorten the navigable route two collateral cuts were made between the Lea and the Thames and the Regent's Canal. The first was the *Limehouse Cut* (TQ 383 823 to TQ 360 809), 1¾ miles long, between the Lea at Bromley-by-Bow to the Thames at Limehouse to avoid navigating round the Isle of Dogs. John Smeaton proposed the scheme in a report of 1766 and in 1767–70 Thomas Yeoman supervised the construction of the cut. The second link was the *Hertford Union Canal* (TQ 373 849 to TQ 359 832) promoted by Sir George Duckett of Bishop's Stortford. The canal is a mile long and was built under an Act of 1824. Completed in 1830, it is still often referred to as 'Duckett's Canal'.

2. Waltham Abbey Powdermill

HEW 1442
TL 377 025 to TQ 378 988

The Lea Valley was one of the most important areas for the manufacture of gunpowder in Britain. Eight sites were in production during the seventeenth century, but Waltham Abbey was to become by far the most important. The raw materials are saltpetre (75%), charcoal (15%) and sulphur (10%). Powder used for mining, quarrying and civil engineering purposes had the proportions saltpetre (70%), charcoal (15%) and sulphur (15%). Charcoal was readily available on site, but the sulphur was imported from Sicily and the saltpetre from Scinde in India. Water power was used to incorporate the ingredients, and the river and navigation were used to transport the finished product. Gunpowder factories are characterised by remote sites with widespread buildings to reduce accidental blast damage. The works was in production in the mid-1660s and was in private hands until it was sold to the Crown in 1787, when it became known as the *Royal Gunpowder Factory*, which was only formally closed on 28 July 1945. It continued as a Government research establishment until the Ministry of Defence decommissioned the site in 1991.

The factory is divided into two sites: the original north site (253 acres) and the south (212 acres), which was acquired in 1885 for a new guncotton factory. The site contains some 300 structures, including 21 listed buildings and is a Scheduled Ancient Monument. Today the works are managed by the Waltham Abbey Trust Company, which is actively developing the powdermill as a public visitor attraction. This is the most significant heritage site in Britain relating to gunpowder manufacture.

BUCHANAN, B. J. Waltham Abbey Royal Gunpowder Mills: The Old Establishment. *Trans. Newcomen Soc.*, 1998–99, **70**, No. 2, 221–50.

RCHME. *The Royal Gunpowder Factory, Waltham Abbey, Essex*. Royal Commission on the Historical Monuments of England, Swindon, Survey, 1993.

CROCKER, G. *Gunpowder Mills Gazetteer*. Society for the Protection of Ancient Buildings, London, Occasional Publication 2, 1988.

GRAY, E., March, H. and McLaren, M. A short history of gunpowder and the role of charcoal in its manufacture. *J. Mater. Sci.*, 1982, **17**, 3385–400.

3. Royal Small Arms Factory, Enfield

HEW 2242
TQ 378 994 to
TQ 379 987

The factory is on an island site of 94 acres between the River Lee Navigation and its flood relief channel. In 1804 the Board of Ordnance issued a Warrant for small arms to be made at the Tower of London, and in the same year instructed Capt. John By, of the Royal Engineers, to establish an arms factory on the River Lea at Enfield. Nothing happened at once, but land was subsequently purchased and in August 1811 John Rennie gauged the river and estimated the power available and suggested the layout of the works. The manufacture of small arms continued on a relatively small scale of production until, stimulated by the Crimean War, the works were greatly developed in 1854–59. The factory was re-equipped with American machine tools from Springfield, New England, and complete firearms were manufactured, namely the 'Lock, Stock and Barrel'. In the 1880s it was said that the machinery here was probably 'the most perfect of any gun-making establishment, whether private or Government, at home or abroad'.[1]

The historic core of the works occupies about 30 acres and arms production ceased in 1988. The site is being developed for a range of public and private purposes. The location of the site accounts for the name of the factory's most famous product, the Lee-Enfield rifle.

[1]WALFORD, E. *Greater London*, vol. 1. Cassell, London, 1880, 399.

4. River Westbourne and the Serpentine Lake

HEW 2243

This river has its origins in a number of small streams in West Hampstead and becomes one river in Paddington. Immediately after crossing the *Bayswater Road* the river enters Hyde Park and joins the Thames in the grounds of Chelsea Hospital. In 1726 James Horne was engaged to direct engineering works in the deer park of Kensington Palace. The scheme was planned in two phases: the excavation for the *Round Pond* close to the Palace, and the construction of a dam to produce *the Serpentine* lake

Sloane Square Station, aqueduct carrying River

DENIS SMITH

(TQ 267 807 to TQ 278 800). These works were under the control of the office of HM Woods and Forests

The first stage, undertaken 1726–28, involved excavating nearly 60 000 cu. yd. to form the 7 acre round basin, and at the same time a temporary dam on the site of the present bridge. For the second stage Horne planned the main dam in July 1730; work began in September and the dam was completed in May 1731. The earth dam stood 17 ft above ground level with 3 ft freeboard; it was about 515 ft long with a 60 ft crest forming a terrace which was raised a further foot in 1827.

The outfall from Hyde Park flows south towards the Thames, and when the Metropolitan District Railway built Sloane Square station it became necessary to carry the River Westbourne over the railway platforms in a cast-iron pipe culvert. The overhead culvert is still visible from the station platforms today.

SKEMPTON, SIR A. W. (ed.). James Horne. *Biographical Dictionary of Civil Engineers in Great Britain and Ireland: 1500–1830*. Thomas Telford, London, 2001.

5. The Serpentine Bridge

HEW 2243
TQ 269 802

The Serpentine lake in Hyde Park was formed by damming the River Westbourne, and flooding part of the river's valley. The bridge over the Serpentine was commissioned by the Office of Woods, Forests and Land Revenue, the government department in charge of national property. The site for the bridge was at an existing dam that separated Kensington Gardens from Hyde Park. George Rennie, in association with his father, was appointed engineer, and on 7 August 1824 the Commissioners accepted their design and estimate. The specification was submitted on 22 September 1824. The foundations were constructed on 9 in. diameter timber piles 15 ft long, at 4 ft centres. There are five segmental arches, each of 40 ft span, with a rise of 4 ft 6 in. A small land arch at each end allows passage along the banks of the Serpentine. The specification stipulated that Pozzolana mortar should be used for the exterior part of the bridge and extending 2 ft inwards, and lime mortar for the whole of the interior. Tenders were received in November 1824 and Jolliffe and Banks were appointed

DENIS SMITH

Serpentine Bridge

contractors. This elegant bridge was opened in 1826 and remains in use today.

6. St. James's Park Footbridge

HEW 2378
TQ 295 798

The park is one of the oldest in London, having been formed during the reign of Henry VIII, and occupying about 87 acres. Charles II employed the French architect Le Notre (who had designed the gardens at Versailles) to redesign the park, and he created the lake from the existing chain of small ponds. During the Georgian and Regency periods the lake was known as a 'Canal' running from Buckingham Palace to Whitehall. The park as we know it today was created by Nash in 1827.

The elegant footbridge over the lake in St. James's Park, Westminster, replaced a mid-nineteenth century suspension bridge designed by J. M. Rendel. Built in 1957, it is an early London example of the use of the Gifford–Udall system of prestressing concrete. There are three spans, of 35 ft, 70 ft and 35 ft, which, in service, are continuous over the supporting piers. The bridge has an overall width of 13 ft 6 in. Recognising that such a slender structure might be subject to noticeable oscillation under live load, the designers took care to ensure that the

DENIS SMITH

St. James's Park footbridge

frequency and amplitude of any vibration would be within acceptable limits.

The Structural Engineering Branch of the Ministry of Works designed the bridge and Higgs and Hill Ltd. were the contractors.

WALLEY, F. St. James's Park Footbridge. *Min. Proc. Instn Civ. Engrs*, 1959, **12**, 217–222.

7. North Circular Road Aqueduct

HEW 2245
TQ 193 836

The Grand Junction Canal (HEW 1718) was built under an Act of 1793, and runs from Braunston to London, reaching Brentford in west London in 1801. Another Act of 1795 authorised a branch canal from Bull's Bridge for a level waterway of 13½ miles to a basin in Paddington. Basin warehouses were built at Paddington, which soon became a busy terminus. The Grand Junction Canal later became part of the Grand Union Canal.

When the North Circular Road was planned in the 1920s a problem arose where the east–west aligned road met the embankment of the north–south aligned Grand Union Canal on its route to Paddington. An aqueduct to carry the canal over the new road was the only solution, and this had to be constructed without interrupting the canal traffic. The design was undertaken by Alfred Dryland, the Middlesex County Engineer, and his Resident Engineer was W. Casson. A design in reinforced concrete for an aqueduct with a central support between the two dual-carriageway roads was made giving two equal spans of 37 ft. The central support comprised a ground beam (transverse to the canal) and ten square columns, of which six supported the canal bed, two on one side supported the 8 ft wide footpath and two supported the 12 ft wide carriageway on the other side. The bed slab was 6 in. thick supported by six beams (longitudinal to the canal) each 2 ft 3 in. deep (below the slab) and 18 in. wide. The canal has a 50 ft wide water surface and the aqueduct had a watertight lining of blue brick laid in asphalt. The main contractor was Roads & Public Works Ltd., who built the aqueduct in two longitudinal halves in steel cofferdams to maintain canal use during construction, which was completed in December 1933.

In the 1990s the North Circular Road was re-engineered to provide dual three-lane carriageways and the problem of replacing the earlier aqueduct with a single wide-span structure had to be faced. The new aqueduct structure provides two 5 m wide channels for the canal, separated by a 5 m wide hollow concrete box. The structure was built alongside the canal on PTFE-coated sledges and then moved into position by jacks over the Easter weekend of 1993. The total weight of the structure is 3000 tonnes, carrying 1000 tonnes of water over a 52 m span. The consulting engineers were Mott MacDonald, the contractors were Balfour Beatty and the client was the Department of Transport.

Aqueduct over the North Circular Road. *Concrete Construction Engrng*, 1933, **28**, No. 12, 693–98.

Construction News, 25 March 1993, 16.

8. The Regent's Canal

HEW 1718
TQ 267 815 to
TQ 363 810

The idea of building a canal from the Paddington Basin of the Grand Junction Canal to the Thames at Limehouse was first explored in 1802 by a promoter named Thomas Homer who had consulted John Rennie.

Rennie proposed a central London route to the East End, but this proved too costly. It was not until 1811, when Homer approached the architect John Nash, who considered that the canal would enhance his *Marylebone Park* (later to be called *Regent's Park*), that progress was made. A company was formed under an Act of July 1812, and in August James Morgan (an Assistant in Nash's office) was appointed 'Engineer, Architect, and Land Surveyor' to the company. Work began, but in 1817 the company ran short of money and Thomas Telford was asked to report to the Exchequer Loan Commissioners, after which successive loans up to £250 000 were made to complete the work. A distinctive feature of the canal is its use of pairs of parallel locks, 78 ft long and 14 ft wide, which were designed to reduce the loss of water from the summit level. Two major tunnels were necessary on the route. The canal is 8½ miles long and is on one level from Paddington to Hampstead Road Locks. From there it falls 90 ft through a further 12 locks to the Limehouse Basin (Regent's Canal Dock). The completed canal to Limehouse was opened on 1 August 1820. The Regent's Canal Dock was built under an Act of 1819 and occupies 4½ acres. James Morgan's assistant engineer here was James Tate and the excavation was undertaken by Hugh McIntosh. In 1928 the Regent's Canal amalgamated with the Grand Junction and other canals to form the Grand Junction Canal Company. The Regent's Canal was much used for carrying coal to the adjacent gasworks and for conveying building materials so important to the development of London. The company agreed to build a large dock at Limehouse, which was capable of accommodating sea-going vessels, and this was also opened in 1820. Canal traffic entered the dock, now mainly known as *Limehouse Basin*, under a bridge carrying the Commercial Road and under the viaduct of the London & Blackwall Railway. The dock underwent various alterations, reaching its final state in 1968. The Limehouse Cut from the River Lea also enters the basin. In 1830 a connection between the Regent's Canal and the Lea Navigation was opened. It is known as the Hertford Union Canal (HEW 1718) (TQ 358 832 to TQ 373 849) and this waterway, sometimes known as Duckett's Canal, is 1½ miles long.

Commercial traffic ceased during the 1960s, and in the mid-1970s one of each pair of locks was converted to a flood weir.

SPENCER, H. *London's Canal: An Illustrated History of the Regent's Canal.* Putnam, London, 1961, reprinted 1976.

9. Maida Hill and Islington Tunnels

HEW 2244

Both tunnels were designed by James Morgan, Engineer to the Company. The Maida Hill tunnel is constructed in brick, is 370 yd long, and carries the Regent's Canal under the Edgware Road, part of the Holyhead Road (A5). The contractor for the tunnel was Daniel Pritchard.

The Islington Tunnel (TQ 309 834 to 317 833), also of brick, carries the canal under Islington Hill and the New River at a depth of 60 ft below the summit. The tunnel is 960 yd long, 19 ft 6 in. high and 17 ft 6 in. wide. The contractor was Daniel Pritchard, whilst Hugh McIntosh undertook most of the earthwork. It was built without a towpath and horses were led over the hill to rejoin their boats, which were at first legged through by teams of men employed by the canal company. A steam tug, which hauled itself along a chain laid through the

Regent's Canal, Islington Tunnel portal

WENDIE TEPPETT

tunnel, was introduced in 1826. Similar tugs continued in use until the 1930s.

10. Macclesfield Bridge, Regent's Park

HEW 2246
TQ 275 833

This interesting structure, designed by James Morgan, Engineer to the Regent's Canal Company, was built between 1815 and 1816, and carries *Avenue Road* over the Regent's Canal in to the Outer Circle of Regent's Park. It is named after the Earl of Macclesfield, who was Chairman of the Canal Company. There are three brick arches carried on cast-iron fluted Doric columns, with square capitals. The north arch spans the canal towpath. Circular features in the spandrels above the springings add to the distinctive character of the bridge. Two rows of iron columns, five in each row, are spanned by four transverse brick arches between the capitals and a wrought-iron tie rod contains the thrust in the outer arches. The ironwork was supplied by the Coalbrookdale Company, and the edge of the capitals are marked 'Coalbrookdale' in relief lettering. The bridge is also known as *Blow-up Bridge* following an explosion in October 1874 when a

Macclesfield Bridge, Regent's Park

WENDIE TEPPETT

steam tug with a train of five barges, one of which was carrying 5 tons of gunpowder, exploded, demolishing the bridge and causing great damage in the neighbourhood. The bridge was rebuilt but there is still evidence of blast damage to the cast-iron columns.

11. The Brent and Ruislip Reservoirs

HEW 2247

The Regent's Canal Company encountered problems in keeping water in their Summit Level at Camden Town. To alleviate the problem the Regent's Canal Company built two high-level feeder reservoirs in north-west London. The Company built a huge reservoir on the river Brent near Kingsbury (TQ 217 873), formed by an earth dam and a massive masonry semi-circular overspill weir at the west end of the reservoir. The engineer was James Morgan, and by mid-January 1834 he had staked out the ground for a 47-acre reservoir holding 1400 locks of water. In September 1834 Morgan submitted plans for a 69 acre reservoir holding 2100 locks, which Morgan calculated it would fill three times a year. Work began in November 1834, the contractor being William Hoof of Hammersmith. By the end of December 1834 the foundations and the main and side walls of the weir in the dam were complete. In January 1835 Morgan suggested raising the embankment to contain a further 1000 locks. The reservoir was filled for the first time in early November 1835. In the September Morgan had resigned as Engineer to the Company, but kept an active interest in the scheme. In December 1836 he proposed enlarging the capacity by raising the dam and buying more land. In April 1837 the Company decided to acquire a further 66 acres (more than doubling the capacity) and the enlarged reservoir was filled on 29 November 1838.

In 1933 the reservoir was inspected by W. J. E. Binnie under the Reservoir (Safety Provisions) Act of 1930 who found that the practice of using stop planks to raise the water level was unsafe. Major work was undertaken by Holst & Co. Ltd., including the installation of five cast-iron siphons. During the Second World War the water level was lowered as a safety precaution, and now the water is kept at 3½ ft below the crest of the overflow weir.

The reservoir was originally known as *Kingsbury Reservoir*, but is now better known as *The Welsh Harp* reservoir. From its opening through to the present it has been a popular site for leisure pursuits.

The Ruislip Park Reservoir (TQ 087 893), again formed by an earth dam, is in the grounds of *Ruislip Park* and extends to 80 acres, and is now much used for water sports. The reservoir was built by the Grand Junction Canal Company in 1810.

FAULKNER, A. H. The Welsh Harp Reservoir. *J. Railway Canal Hist. Soc.*, 2000, **33**, Pt. 4, No. 175, 262–72.

Grosvenor Canal, lock and basin

DENIS SMITH

12. Grosvenor Canal, Pimlico

HEW 2241
TQ 289 789 to
TQ 287 780

The commercial origins of this canal are closely related to those of the Chelsea Waterworks. The waterworks was formed under an Act of 1721 (7 Geo.1 c.26). In 1727 the company leased a plot of land on the Grosvenor estate, close to the Thames at Pimlico. The plan was to pump water into reservoirs in the area of Hyde Park. Cuts were made from the Thames for the water supply, but it soon became clear that they had potential as a northerly navigable waterway from the river. The canal comprised twin cuts terminating at Pimlico Wharf, which was later enlarged to form the Grosvenor Basin. This canal was to share a similar fate to that of the Kensington Canal, namely to be made part of a railway route. The *Victoria Station and Pimlico Railway* obtained their Act in 1858 and the Grosvenor Basin became the site of Victoria Station. A short length of the lower part of the canal remains in use today by the City of Westminster, where a lock and basin are used to load urban refuse into barges to be deposited downstream as landfill on the Thames estuary marshes.

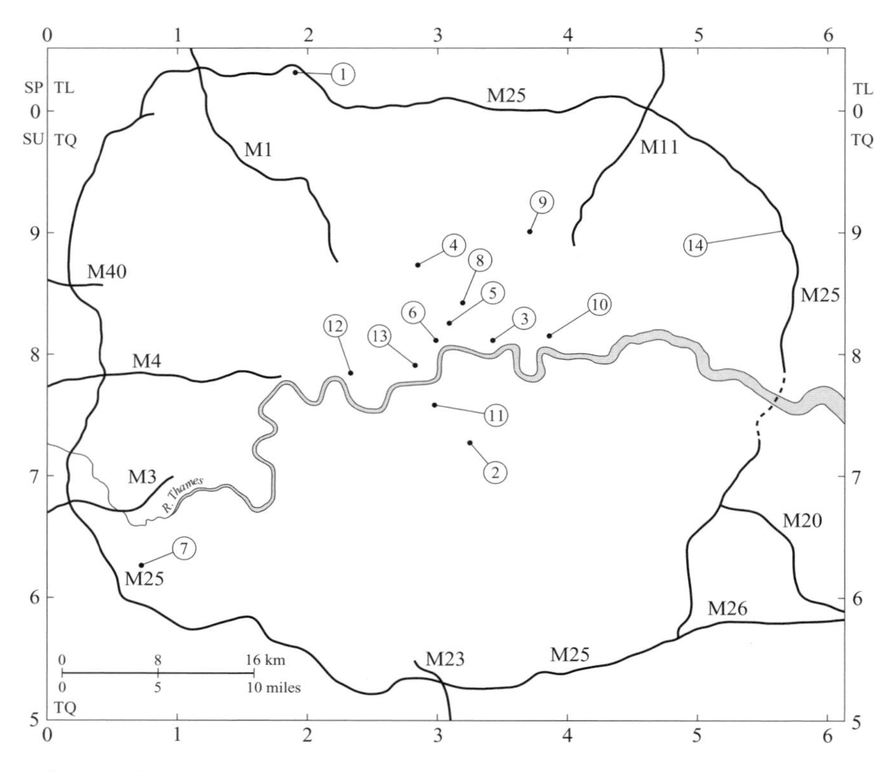

1. Seven Arch Bridge, London Colney, Hertfordshire
2. Dulwich Toll House and Barrier
3. The Commercial Road
4. Highgate Archway
5. Holborn Viaduct
6. Kingsway Tram Subway
7. Brooklands Motor Racing Track
8. New North Road and Bridge
9. North Circular Road and Wadham Road Viaduct
10. Silvertown Way
11. Stockwell Bus Garage
12. Hammersmith Flyover
13. Park Lane Improvement Scheme
14. The London Orbital Motorway (M25)

5. Roads and Road Transport

From the earliest origins of the settlement of London the establishment of a rudimentary system of trackways was a necessity. These early routes have influenced the subsequent planning and development of the capital. A study of consecutive maps is essential in understanding both the development of new roads and, indeed, the surprising survival of the alignment of ancient roads. The sites of the old gates of London provide evidence of the Roman roads leading out of London. *Watling Street* (HEW 1873) and *Ermine Street* were the principal roads leaving the centre of the City. From Newgate a road led to the area now known as Marble Arch, where Watling Street diverted north-west to Verulamium (St. Albans) whilst the west road ran on to Silchester. The main road to the north, *Ermine Street*, left London at Bishopsgate, and the road to Colchester via Aldgate. Archaeologists are still finding evidence of Roman roads, and at *Bow Lane/Watling Street* (TQ 324 810) in the City of London, part of a Roman road was recently discovered that is considered to be a precursor of the present *Watling Street*.[1] It is significant that the Romans first settled on the north bank of the Thames rather than tackle the problems posed by the marshy ground on the south bank.

The financing and management of road building had evolved slowly. The Turnpike Trust Act of 1663 was the first to embrace the idea that road users should pay for the construction and maintenance of roads. The 1663 Act authorised gates on the road north, the nearest to London being at Wades Mill in Hertfordshire. No further Turnpike Acts were passed until 1696, when the road from London to Harwich was promoted. The importance of the turnpike system was that it employed surveyors who were to be the origins of a professional body of men dedicated to a career in road design, construction and maintenance. In 1815 the Government formed a body of Commissioners to administer the construction and improvements of the road from London to Holyhead, with Thomas Telford as Engineer. Early nineteenth-century turnpike road building in north London included *The Commercial Road* (1802), *New North Road* (1812), *Archway Road* (1813), *Caledonian Road* (1826) and *Finchley Road* (1826–55). In 1826 an Act of Parliament amalgamated the 14 Turnpike Trusts north of the Thames under Government Commissioners, whereas in south London there were still 15 separate Turnpike Trusts in 1831. Many other Turnpike Trusts were established in the Greater London area, and by the mid-nineteenth century toll gates and toll houses were a common feature in the townscape. A major development in the early nineteenth century was the construction of the 'New Road' now known as the *Euston Road*.

The formation of the *Metropolitan Board of Works* in 1856 was a major factor in the development of new road transport routes in London. The Metropolitan Board of Works had responsibility for 117 square miles of London and for the first time it was possible to form a capital-wide strategy in town planning. The Metropolitan Board of Works had powers of compulsory purchase and this led to the demolition of buildings to form new streets and major roads. Under their Engineer, Sir Joseph Bazalgette, the Metropolitan Board of Works constructed several new streets. Notable examples, with the dates of opening, are *Southwark Street* (1864), *Queen Victoria Street* (1871), *Northumberland Avenue* (1876), *Shaftesbury Avenue* (1886) and *Charing Cross Road* (1887). In these main thoroughfares Bazalgette provided service ducts to minimise subsequent digging up of the roads. In addition, the opportunity was seized of providing additional roads on his new Victoria, Albert and Chelsea Embankments. Later in the nineteenth century tolls were progressively abolished, and in 1888 the Local Government Act gave the new County Councils powers as Road Authorities.

The first horse-drawn omnibus, developed by George Shillibeer, appeared on the streets of London in 1829, and the four-wheeled Hansom Cab followed in 1834.

Horse-drawn tramways became an important part of London's public transport, and in 1870 the Tramways Act set out regulations for their construction. The Act required that all tramways should be built to the standard railway gauge. Various companies and Local Authorities became involved, and on its formation in 1889 the London County Council was given powers to purchase tramway companies. Between 1892 and 1903 the London County Council gradually acquired all the London undertakings, all of which were horse-drawn and not fully interconnected. The London County Council unified the whole system, converted it to electricity, built the Kingsway subway and Greenwich power station, and carried the tramways over certain Thames bridges and along the embankment. In 1933 the London County Council tramways were transferred to the newly formed London Passenger Transport Board which was replaced by the London Transport Executive on 1 January 1948.

The advent of the motor vehicle at the end of the nineteenth century provided a major incentive to improve roads and led to taxation to finance the work. The 1903 Motor Act raised the permissible speed to 20 miles per hour. In 1909 the Chancellor of the Exchequer levied a tax on pleasure motor vehicles and a duty on petrol. In the same year the Road Improvements Act was passed and a Government Department was established administer the funds. It became known as 'The Road Board' and became operational in 1911, and two years later was asked to classify roads into three classes, the first two of which would receive substantial maintenance grants.

Speaking in June 1911 D. Ellison, MICE, argued for more experimental research on road construction and maintenance saying:

> The phenomenal growth of rapid transit locomotion and the advent of an entirely new class of traffic, with its attendant problems and demands, coupled with the great and increasing expenditure, endow the question with great importance, and it is a matter which British Engineers should strive to further to the utmost extent of their opportunities. With a development of our knowledge of the properties of materials, undreamt of at the present moment, may be possibly obtained.[2]

An Act of 1919 established the Ministry of Transport, and the Road Board was wound up in 1920. The Board had initiated only two major arterial road schemes for London, the *Great West Road* and the *Croydon Bypass*, and neither was completed during their term of office.

Traditionally, road bridges were constructed either of timber or masonry, but during the nineteenth century cast iron, wrought iron and steel were increasingly used. From the late 1920s and through the 1930s, during the period of arterial road building, and in the 1960s motorway building programme, reinforced concrete became a popular material for road bridges. The historical engineering works described here are arranged chronologically by date of construction.

[1]*Greater London Archaeology Advisory Service, Quarterly Review*, April–June 2000. English Heritage.

[2]ELLISON, D. Collection of data for standardisation of road construction and maintenance. *Proc. Instn Mun. County Engrs*, 1911–12, **38**, 114–17.

MAYBURY, SIR H. P. Presidential address. *Min. Proc. Instn Civ. Engrs*, 1933–34, **237**, Pt. 1, 1–23.

ALBERT, W. *The Turnpike Road System in England 1663–1840*. Cambridge University Press, Cambridge, 1972.

RAYFIELD, F. A. The planning of ring roads, with special reference to London. *Min. Proc. Instn Civ. Engrs*, 1956, **5**, Pt. 2, 99.

1. Seven Arch Bridge, London Colney, Hertfordshire

HEW 2353
TL 182 037

There has long been a recognised road route out of London to the north-west via Holloway, Whetstone, Barnet, South Mimms and St. Albans, and this was described by Ogilby in 1675 as part of the route to Holyhead. The road was one of the first to be turnpiked, with the section from St. Albans to South Mimms receiving its Act in 1715. (The section from Highgate to Barnet was turnpiked in 1711; see HEW 1214.) The section across the Colne Valley was improved in the 1760s, when the present Seven Arch Bridge at London Colney was constructed by a local builder (*c.* 1772–75). Approached on an embankment on a gentle course, it comprises seven semi-circular brick arches of 5, 8, 11, 13, 11, 8 and 5 ft span. The three middle spans are founded on inverted arches in the river. In the 1820s the Turnpike came under the auspices of the Holyhead Road Commissioners and Thomas Telford recommended improvements to the approaches to the bridge. At that time the Engineer to the Trust was James McAdam, and in the late 1820s there was something of a dispute concerning the condition of the road and the McAdam system of construction, which was frowned upon by Telford and his supporters.

Seven Arch Bridge, London Colney

DENIS SMITH

The road was classified as part of the A6, and in 1948 the original brick parapets were replaced by metal railings. In the late 1950s unsympathetic repairs were carried out, including new footways supported on piers attached to the spandrel walls. The road was bypassed in the 1960s by the A6 trunk road improvements, and in 1997, as the result of an assessment of the bridge, it was decided to strengthen the structure, at the same time removing many of the previous unattractive repairs. A reinforced-concrete relieving slab was supported on internal piers founded on the original substructure, and the footway extensions were removed and replaced by reinforced-concrete cantilevered footways integral with the relieving slab. This work has permitted the original arches to be viewed once more. New Portland stone was used to replace damaged stone copings, groins and cantilevers. Handmade bricks and hydraulic lime were used elsewhere. The repairs, designed by L. G. Mouchel & Partners and carried out by Wrekin Construction Ltd., were completed in July 1998.

LANSBERRY, H. C. F. James McAdam and the St. Albans turnpike. *J. Transport Hist.*, 1965, **7**, Pt. 2, 121–23.

JERVOISE, E. *The Ancient Bridges of Mid and Earlier England*. Architectural Press, London, 1932.

Parliamentary Papers Reports: (1824–50), HC126, *Commissioners ... for the further Improvement of the Road from London to Holyhead*; T. Telford (1820) *Report ... upon the State of the Road between London and Shrewsbury*.

OGILBY, J. *Britannia: or an Illustration of the Kingdom of England*. London, London, 1675.

2. Dulwich Toll House and Barrier

HEW 1390
TQ 335 726

Dulwich College, the south London public school, was founded by George Alleyn in 1613. In 1787 John Morgan, Lord of the Manor of Penge, built a house, together with a substantial access road, on land leased from the Governors of the College. In 1789 the College approved the construction of a toll house and gate so that Morgan could accumulate funds from other road users to maintain his road. The road attracted a large amount of traffic, as it was a direct route between Camberwell and Penge. In 1802 the lease of two fields and the access road was assigned to a John Scott, who in 1809 surrendered it to the College.

Dulwich Toll House

DENIS SMITH

The road was originally named *Penge Road*, but was renamed *College Road* after the building of the new college in 1870. By the end of the nineteenth century tolls were removed from all roads in London, except at Dulwich.

The toll house stands at the junction of *College Road* and *Grange Lane* on the college estate. The present toll house is built of brick and tile with a timber porch, and the toll barrier is of painted post and rail. There is no longer a toll gate, the traffic passing through a 9 ft gap in the timber fence crossing *College Road*.

In 1871 it was said that the curriculum at the college, included 'mathematics, the natural sciences, chemistry, the principles of civil engineering, and all the branches of a liberal education'.[1]

[1]THORNE, J. *Handbook to the Environs of London*. John Murray, London, 1876, 156.

3. The Commercial Road

HEW 2257
TQ 340 813 to TQ 370 811

This turnpike road was promoted by trustees to carry the heavy goods wagon traffic from the West India Docks to the City. Ralph Walker was first approached, but his arrangement with William Jessop at the West India Docks precluded him taking on other engineering work. Ralph suggested his young nephew, James Walker, who was appointed Engineer. An Act was obtained in 1802 and the road was built as a common paved turnpike road. By the 1820s there were complaints that the heavy traffic was destroying the road surface, and a group of private promoters planned to lay an iron railway along the road. The road trustees opposed this scheme, but sought to improve the road. James Walker was again consulted, and he suggested a road with lanes dedicated to particular types of traffic. An Act was obtained in 1828 and a stone tramway was laid. In March 1829 Walker carried out some interesting experiments with heavily loaded horse-drawn wagons and measured the tractive effort with a dynamometer, showing that, on the level, 12 lb force would move 1 ton, and reporting that 'This friction is not more than upon the best constructed edge-railway'. Walker remained Engineer to the Trustees of this important road until his death in 1862.

4. Highgate Archway

HEW 2258
TQ 291 874

For centuries the main road from London to the North West passed through Highgate Village, standing on a steep hill 400 ft high. Highgate Hill became a severe problem with the introduction, and growth, of stage-coach traffic. In 1808 Robert Vazie produced a design for an archway (or tunnel) under the hill, but nothing was done at that time. An Act was obtained in 1810, and in April 1811 John Rennie was consulted and

DENIS SMITH

Highgate Archway

suggested an open cutting rather than a tunnel. Nevertheless, tunnelling was started, and after a length of only 130 yd was completed the whole structure collapsed on 13 April 1812. A new design was produced, comprising an open cutting with a masonry viaduct with a central archway 18 ft wide and 36 ft high surmounted by an upper viaduct with three central semi-circular openings carrying *Hornsey Lane* over the *Archway Road*. The foundation stone was laid in October 1812 and the road was opened in August 1813. In 1829 the Commissioners for the Holyhead Road took temporary control of the road when Telford and Macneill greatly improved the foundation of the road. Tolls were abolished on the Archway Road in 1876. By 1871 horse-drawn trams had reached the foot of the hill and in May 1884 a cable tramway, the first of its kind in Europe, was opened up Highgate Hill. The 3 ft 6 in. gauge tramway was 3780 ft long, of which 2700 ft was double track, the remaining 1080 ft being a single line with two passing places. Following a serious accident the line was closed in 1894. The tramway was bought by a new operator, the line was reconstructed and new plant installed and re-opened in April 1897.

In the late nineteenth century the London County Council had to consider replacing the masonry structure, a work they described as an 'exceptional undertaking'. The necessary powers were obtained under the London County Council (Improvements) Act of 1894, and in December they estimated the cost at £28 000. In September 1896 the London County Council, with Alexander Binnie as Engineer, authorised the work comprising a high-level steel arch structure of 120 ft span, and in July 1897 accepted the tender of Charles Wall for £25 626. The new bridge was opened on 28 July 1900.

MORRIS, S. and MASON, T. *Gateway to the City: The Archway Story*. Hornsey Historical Society, London, July 2000.

5. Holborn Viaduct

HEW 561
TQ 316 816

During the eighteenth century the Fleet river was increasingly culverted by arching it over. By 1732 it had been covered down to Holborn Bridge, by 1742 as far as Fleet Street, and by 1769 from Fleet Street to the Thames, leading to Robert Mylne's Blackfriars Bridge opened in that year. In 1863 the Corporation of the City of London obtained the London Coal and Wine duties Act to finance improvements to the Holborn Valley, and an Act for a new Blackfriars Bridge. The Holborn Valley scheme included the building of a viaduct to replace old Holborn Bridge. William Haywood, the City Engineer, was appointed to design the scheme in June 1864. Work on

Holborn Viaduct

WENDIE TEPPETT

site began in 1867 and the viaduct was opened in 1869. The general building contractors were Hill & Keddell, and the contractors for the iron superstructure were Cochrane & Grove of Dudley. The 1400 ft long masonry viaduct is a complex structure designed with subways for a sewer, gas main, telegraph wires, the pneumatic despatch railway and an Edison electric power station. A skew bridge spans the Farringdon Road. The bridge comprises granite piers carrying cast-iron arched soffit girders with corrugated cast-iron decking carrying the road. During 1990–91 repairs to cracks in the cast-iron girders were repaired by the Metalock system and the original corrugated cast-iron decking was replaced with a reinforced-concrete slab.

6. Kingsway Tram Subway

HEW 2259
TQ 307 807 to TQ 307 812

This subway, the first built to ease road congestion in London, was built for the London County Council tramways to provide a through route connecting the tramway systems north and south of the Thames. It is a complex mixture of gradients, alignments, cross-sections and construction techniques, and runs from *Southampton Row* to the *Victoria Embankment* at Waterloo Bridge. Although the scheme was first suggested in 1889, work did not begin until 1904.

At the northern end the ramp is clearly visible, dividing the traffic flows in *Southampton Row*. The incline is 170 ft long and 20 ft wide, with a gradient of 1 in 10. This ramp led down to two cast-iron lined tunnels, 14 ft 5 in. diameter and 255 ft long, after which another incline rose

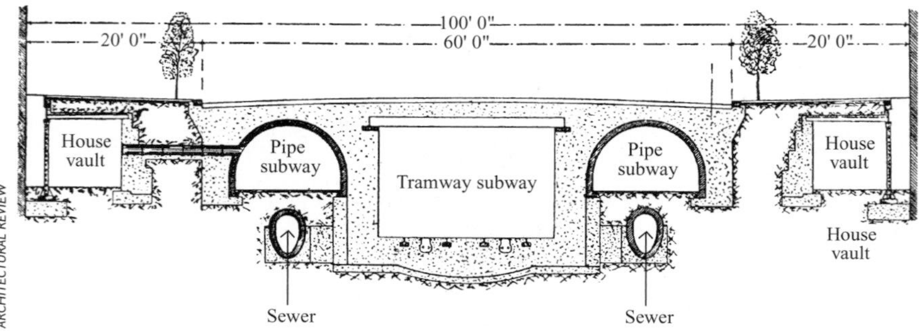

Kingsway Tram Subway

at the same gradient into Holborn station. From Holborn to Aldwych, a distance of 1880 ft, the tracks ran in a single 20 ft wide rectangular tunnel with a roof of steel trough sections, immediately below street level. The tramway as far south as Aldwych was opened in 1906. From Aldwych the tunnel turned westward and dipped in a 1 in 20 gradient passing under *The Strand*. This length, 440 ft long, was of brick arch section. In passing under The Strand the tunnel reverted to twin cast-iron tubes, which curved in plan and ended at the first brick piers of the surface viaduct leading to Rennie's Waterloo Bridge. The final section of the subway involved underpinning 360 ft of this overhead viaduct to allow openings for the 20 ft wide single tunnel. To give access to the Embankment from the tunnel a portal was constructed next to old Waterloo Bridge. Trams first passed through whole subway route in 1908. The Engineer was Maurice Fitzmaurice of the London County Council, and the steelwork was supplied by the Cleveland Bridge Co.

The subway as originally built could only take single-deck trams, and in 1930 major civil engineering work was undertaken to allow for the passage of double-deck trams. The subway was re-opened on 15 January 1931. When the present Waterloo Bridge was built the southern end of the subway was realigned to bring the Embankment portal immediately beneath the bridge. The engineering was supervised at first by Sir Alexander Binnie, and from January 1902 by Maurice Fitzmaurice and assisted by George W. Humphreys, all of the London County Council. The contractor was John Cochrane & Co., and the total cost of the improvement work, including the subway, up to March 1910 was £2 173 359.

The last tram in London ran in July1952, and now part of the subway remains in use by road vehicles as a *Strand* underpass, with access from Waterloo Bridge and emerging in *Kingsway*. The subway is Listed Grade II.

Enlargement of the Kingsway subway of the London County Council tramways. *The Engineer*, 30 January 1931, 136–37.

HUMPHREYS, G. W. The London County Council Holborn-to-Strand improvement, and tramway-subway. *Min. Proc. Instn Civ. Engrs*, 1910–11, **83**, Pt. 1, 21–47.

The Strand to Embankment subway. *The Engineer*, 13 March 1908, 260–61.

7. Brooklands Motor Racing Track

HEW 1428
TQ 070 628

This track was the world's first banked concrete racing track. It was built 1906–07 and was promoted by the civil engineer Hugh Fortesque Locke King. The banked track was also used as a test facility when the speed limit on roads was 20 miles per hour. King built the track on his country estate at Weybridge, and a railway engineer, Col. Capel Holden, designed the layout of the 2¾ mile pear-shaped track, which was steeply banked at the two bends. The engineer was L. G. Mouchel and the contractors were the Yorkshire Hennebique Contracting Co. (Leeds), who had 2000 Yorkshiremen on site.

A reinforced-concrete bridge was designed to carry the track over the River Wey and to provide a passage for floodwaters during periods of heavy rainfall. The bridge was designed for a superload of 112 lbf/sq. ft and a rolling load of 2 tons. The deck of the bridge varied in thickness from 4½ to 6 in. and had a superelevation of 26 ft, with the curvature increasing rapidly towards the outer edge. In 1926 Brooklands staged the first British Grand Prix, and by 1929 cars lapped the circuit at 120 miles per hour.

Another important aspect of the site is its association with aviation and aircraft construction. The A. V. Roe firm was on site in 1908, closely followed by Sopwith and the Vickers Company. Barnes Wallis developed the geodetic fuselage for the Wellington bomber here. Immediately after the Second World War Vickers purchased the site to build new drawing offices and a wind tunnel. From 1945 to 1971 Barnes Wallis used the clubhouse for his experiments. During the 1980s the aviation industry relocated and 40 acres of the site were sold in 1984, of which some 30 acres were set aside for a museum (now open).

TURTON, C. D. An early reinforced-concrete structure: the bridge at Brooklands in Surrey. *Concrete*, August 1969, 313.

Concrete and Concrete Construction, 1907–08, **2**, 240–43.

8. New North Road and Bridge

HEW 2375
TQ 328 830 to
TQ 322 841

The *New North Road*, linking Shoreditch to Highbury, was a new Turnpike Trust road created by Act of Parliament in 1812, and was one of the last turnpikes to be created in the London area. The original brick bridge across the Regent's Canal was replaced by the existing reinforced-concrete bridge, designed by L. G. Mouchel and Partners, 1912–14. The earlier bridge had a span of 25 ft 6 in. and a width of 33 ft. The replacement span of 43 ft 6 in. gave 18 ft more clearance for the canal, and the new bridge was 55 ft 6 in. wide, improving traffic flow. A typical, if late, example of a Hennebique-style beam bridge with parapet girders, it was designed for two lanes of tramway traffic, and is the oldest surviving reinforced-concrete road bridge in the central London area. The outline design was by T. L. Hustler, the Shoreditch Borough Engineer, and the contractors were Higgs & Co. The bridge has been repaired and the original detailed form of the beams has been obscured to some extent by sprayed Gunite.

HUSTLER, T. L. New North Road Bridge over the Regent's Canal, Shoreditch. *Ferroconcrete*, 1914–15, **6**, 228–232.

9. North Circular Road and Wadham Road Viaduct

HEW 2261

The popularity of cycling and the introduction of mass-produced motor vehicles revived interest in the road network in the early twentieth century. The response of Government in the form of the Road Board was to finance a programme of arterial road construction, which was only partially implemented between the Wars. The North Circular Road (A406) was to link the *Green Man*, Leytonstone, in east London to Hanger Lane in west London, bisecting the outer London Boroughs. It would provide continuous road links from north Woolwich to Kew Bridge, linking the main routes out of London and bypassing the central area. The need for the road had been identified by the London Arterial Road Conferences of 1913–16, and it was constructed in a series of improvements providing unemployment relief. Major works were designed by leading consulting engineering practices, with Maxwell Ayrton providing architectural advice. The section from Leytonstone, north of Walthamstow and across the Lea Valley through Edmonton was designed by (Sir) E. Owen Williams. Work began in 1920, with Middlesex and Essex County Council engineers responsible for the work on behalf of the Road Board. Many of the features of the original road, including the Lea Valley Viaduct (1924–27) and Navigation Bridge, were swept away in major improvements carried out to the road in the late 1980s, but a small example of Williams' work remains in Wadham Road. Parts of the original road survive here as a slip road for local traffic alongside the new section in a deep cutting.

The Wadham Road Viaduct crosses the Liverpool Street–Chingford railway line, and is distinguished by its raked pier supports, with rough-cast surface treatment. The unusual features are a reminder of Williams' original approach to reinforced concrete.

WELCH, G. Wadham Road Viaduct. *Concrete Construct. Engrng*, 1930, **35**, 107–109.

WELCH, G. Viaduct and bridges at Barking. *Concrete Construct. Engrng*, 1927, **22**, 243–261.

10. Silvertown Way

HEW 2262
TQ 395 815 to
TQ 400 805

By the 1920s traffic on the old *Victoria Dock Road*, taking traffic from the East India Dock Road to the Royal group of docks, was in a chaotic state. The route involved many road crossings, two railway level-crossings, and a narrow swing bridge over the entrance to the Victoria Dock. A major project evolved to alleviate this congestion, comprising a new Barking Road Bridge over the River Lea (a 200 ft steel arch ribbed structure), a mile-long viaduct known as *Silvertown Way*, and a new bridge of 103 ft span over the tidal basin of Victoria Dock. In addition, the *Silvertown Bypass*, close to Silvertown Station, involving a bowstring girder bridge of 109 ft span, completed the project. The consulting engineers to the scheme were Rendel, Palmer & Tritton and the main contractors were Dorman Long. The scheme involved the demolition of 500 houses, the displacement of 3600 people and raising the road in front of Canning Town Station by 5 ft. Ground conditions were very poor and thousands of concrete piles were necessary. *Silvertown Way*, one of Britain's early flyover schemes, was of reinforced-concrete column, beam, and slab construction. About half way along the concrete viaduct a three-span steel bridge carries traffic over two roads and a railway. The scheme was formally opened by Leslie Hore-Belisha on 13 September 1934.

11. Stockwell Bus Garage

HEW 2264
TQ 304 707

Tramway services in London ceased in 1952 and their replacement by buses led to the need for additional bus garage accommodation. London Transport met this need by converting some former tram depots and by building several new bus garages. Stockwell Garage, in *Lansdowne Way*, was designed to accommodate 200 buses. A shortage of structural steel influenced the decision to use reinforced concrete for the Stockwell garage. A clear floor area of 73 350 sq. ft was obtained by forming the 393 ft long roof structure as a series of two-hinged reinforced-concrete arched ribs. Spanning 194 ft the ribs are 7 ft deep at the crown, increasing to 10 ft 6 in at the haunch, which is formed to provide a minimum headroom of 16 ft from ground level. Arched vaults, also in reinforced concrete, span the 42 ft wide spaces between the main ribs and incorporate roof lights 14 ft wide and 140 ft long. The garage was brought partly into use in April 1952 and into full service during the following year. It is now Listed Grade II.

Stockwell Bus Garage

LONDON TRANSPORT MUSEUM

The architects were Adie Button & Partners in association with Thomas Bilbow, Architect to the London Transport Executive. The Structural Engineer was A. E. Beer and the main contractors were Wilson Lovatt & Sons Ltd.

12. Hammersmith Flyover

HEW 2263
TQ 235 784

The Hammersmith Flyover is at the junction of two important traffic routes; the north–south route from Shepherd's Bush to Chelsea and the Embankment, and the east–west route from central London to Heathrow and the West Country. The flyover is a four-lane overpass and it was constructed in response to complaints about traffic problems from as long ago as the late nineteenth century. Its alignment was finally settled in 1935 and parliamentary powers were obtained by both the London and Middlesex County Councils in 1935. The only part of the route where construction was begun before the Second World War was the extension

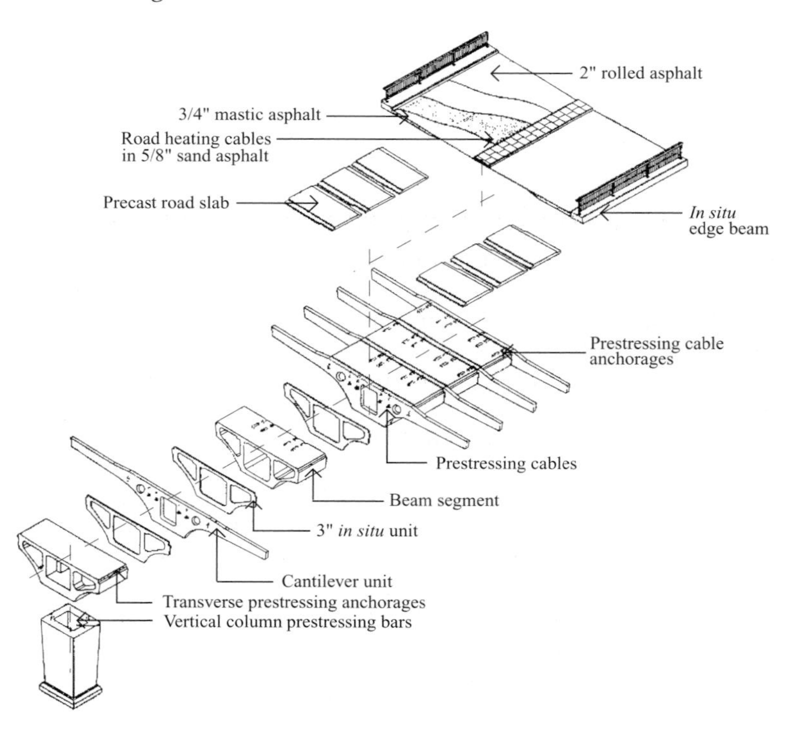

Hammersmith Flyover

of *Cromwell Road*, which was opened to traffic in 1940. Further powers were obtained in 1948 when there was considerable discussion about the relative merits of a tunnel and a flyover. In 1955 work resumed on the *Cromwell Road* extension and in 1956 the Ministry of Transport decided to make a grant towards the cost of the flyover. The constraints of maintaining traffic flow and the cost of property demolition led to the design of a flyover structure supported on single central columns. For the same reasons great attention was paid to the method of erection. The superstructure comprises precast units of hollow rectangular spine beam units, 26 ft wide, with hollow cantilevers on each side. These units, 8 ft 6 in. long, were delivered to site from a casting yard at Heston, and construction proceeded entirely from the western end. The fishbone spine units were prestressed longitudinally. The flyover was completed in 1962 and the final cost was of the order of £1 million.

The consulting engineers were G. Maunsell & Partners, the architect was Mr. Hubert Bennett of the London County Council and the contractor was Marples, Ridgeway & Partners Ltd.

RAWLINSON, SIR J. and STOTT, P. The Hammersmith Flyover. *Min. Proc. Instn Civ. Engrs*, 1962, **23**, 565–599.

13. Park Lane Improvement Scheme

HEW 2355
TQ 284 798

By the end of the Second World War Hyde Park Corner had long been a source of traffic congestion. As early as 1825 Decimus Burton had been employed by the Government to reconsider the layout, and his improvements included new roads and the construction of the triumphal arch in honour of the Duke of Wellington. By 1873 traffic was again at a standstill, and in 1873 a new road in Green Park linked a widened *Piccadilly* and *Grosvenor Place*.

In 1958 the London County Council obtained an Act giving powers for a major reconstruction at the Hyde Park Corner and Marble Arch traffic intersections. By 1960 Hyde Park Corner and Marble Arch were carrying 118 000 and 78 000 vehicles, respectively, in a period of 12 hours.

The London County Council scheme involved the building of a four-lane underpass between *Piccadilly* and *Knightsbridge*, a roundabout at Hyde Park Corner, together with a subsidiary roundabout behind Apsley House, and a roundabout at Marble Arch. In addition, 17 pedestrian subways, totalling more than 1½ miles, were built. These subways have decorative tiled murals.

Work commenced on 9 May 1960 and the scheme was completed by 1964. The work was the occasion for the first use in Britain of ICOS diaphragm walls.

GRANTER, E. Park Lane improvement scheme: design and construction. *Min. Proc. ICE*, 1964, **29**, 293–318.

14. The London Orbital Motorway (M25)

HEW 2332

An outer orbital road for London was proposed in 1913, (see HEW 2261, p. 143), but the concept of an orbital motorway around London first appeared in print in Sir Patrick Abercrombie's visionary *Greater London Plan* of 1944. However, no action was taken at that time. Following the creation of the Greater London Council in 1965 the Greater London Development Plan, which retained Abercrombie's radial and ring structure, was presented for Government approval in 1969. At the end of the 1960s the Government set up a joint planning team with local government to develop an overall regional strategy for the south-east, and in 1976 the Secretary of State approved the Greater London Development Plan, seven years after the Greater London Council had presented it.

During the 1970s the preparation of what was to become London's orbital motorway proceeded, and construction began in 1973. The first short sections were opened in 1975–76. During 1982–83 eighteen miles were completed between the M11 and the A13. By mid-1983 seven separate sections were open, comprising 62 miles of the total 119 miles. The section between Reigate and the M3 was opened during 1983–84. The complete orbital motorway was completed in 1986.

The M25 has a total length of 119 miles and comprises 32 junctions, 234 bridges and three tunnels.

INSTITUTION OF CIVIL ENGINEERS. *20 Years of British Motorways*. ICE Conference, London, 27–28 February 1980.

INSTITUTION OF CIVIL ENGINEERS. *Orbital Motorways*. ICE Conference, Stratford-upon-Avon, 24–26 April 1990.

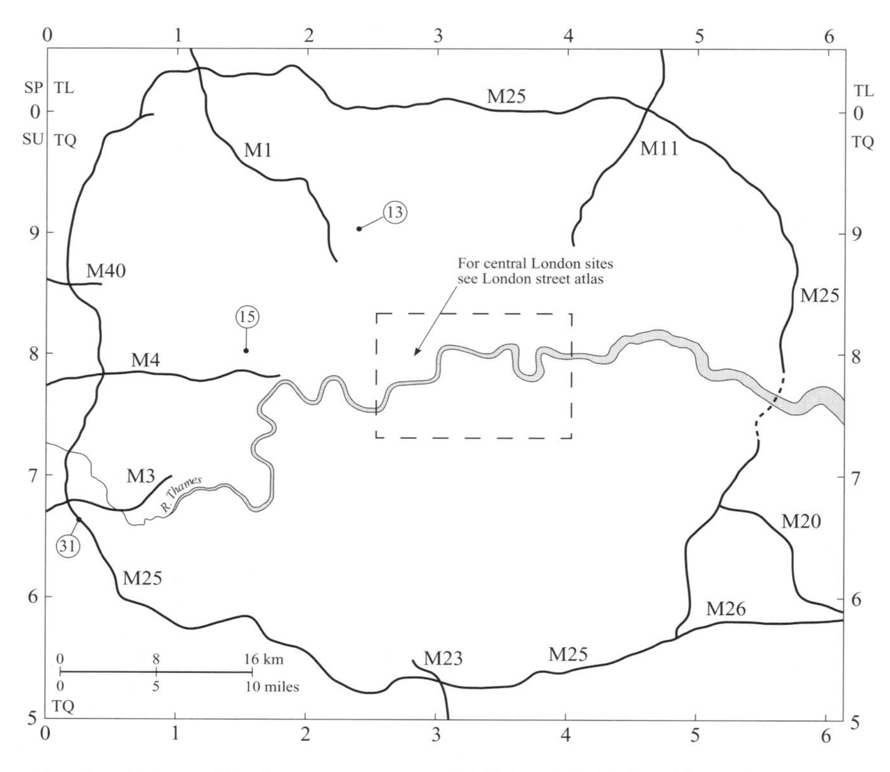

1. London & Greenwich Railway
2. London Bridge Station and Joiner Street Bridge
3. London & Birmingham Railway
4. Primrose Hill Cutting and Tunnel
5. Winding Engine Vaults, Camden Town
6. Round House, Camden
7. London & Blackwall Railway
8. Waterloo Station and International Terminal
9. Great Northern Railway
10. King's Cross Station
11. Gasworks Tunnel under the Regent's Canal
12. The Granary, King's Cross Goods Yard
13. Dollis Road Railway Viaduct
14. Paddington Station
15. Wharncliffe Viaduct, Hanwell
16. Three Bridges, Hanwell
17. Victoria Station
18. Midland Railway
19. St. Pancras Station Train Shed
20. Liverpool Street Station
21. Commercial Street Bridge, Shoreditch
22. Metropolitan Railway
23. Metropolitan District Railway
24. Earl's Court Station
25. London's Early Tube Railways
26. City & South London Railway
27. Post Office Railway
28. Victoria Line
29. Docklands Light Railway
30. Jubilee Line Extension
31. Lyne Railway Bridge

6. Railways

The first public railway in Britain to be authorised by Act of Parliament was the *Surrey Iron Railway*, inaugurated in 1801. The engineer was William Jessop, who designed an iron plateway running along the Wandle Valley from a basin at Wandsworth up to Croydon, using horse-drawn wagons. But in the age of steam it was the *London & Greenwich Railway* which was to be the first to have a terminus in London, at London Bridge, in 1836. During the nineteenth century the various railway companies, arriving from all directions, each sought to have a London terminus. Planning constraints, such as population density on the one hand and high property values on the other prevented their reaching the heart of the capital, and we therefore have an outer ring of railway termini, which originally required a cross-London road journey for through travellers. It is also possible to detect different factors that influenced the pace and pattern of railway promotion north and south of the river. From the north there were considerable mineral and agricultural incentives to encourage railway promotion. The Great Northern Railway brought Yorkshire coal and Fenland farm produce to London and with the Midland Railway there was coal and Burton-on-Trent beer, whereas the revenues of railways arriving from the south and west were largely derived from passenger traffic. The inner London road congestion between the termini led eventually to the construction of the world's first underground railway system. One consequence of the arrival of the railways in London was the compulsory purchase of property, leading to a great deal of slum clearance. The development of the electric telegraph in Britain was pioneered in London when in 1837 the patentees, Cooke and Wheatstone, sought to have it adopted on the London & Birmingham Railway to communicate from Euston Station with the stationary winding engines at the top of the incline at Camden. But the first practical application of the telegraph system was on the Great Western Railway between Paddington and Hanwell in 1839, and it was operational to Slough by 1843. Another pioneering work in London in connection with the railway was the experiment undertaken on atmospheric railway traction. In 1840 the developers of the system, Clegg and Samuda, leased a half-mile of track on the Birmingham, Bristol & Thames Junction Railway

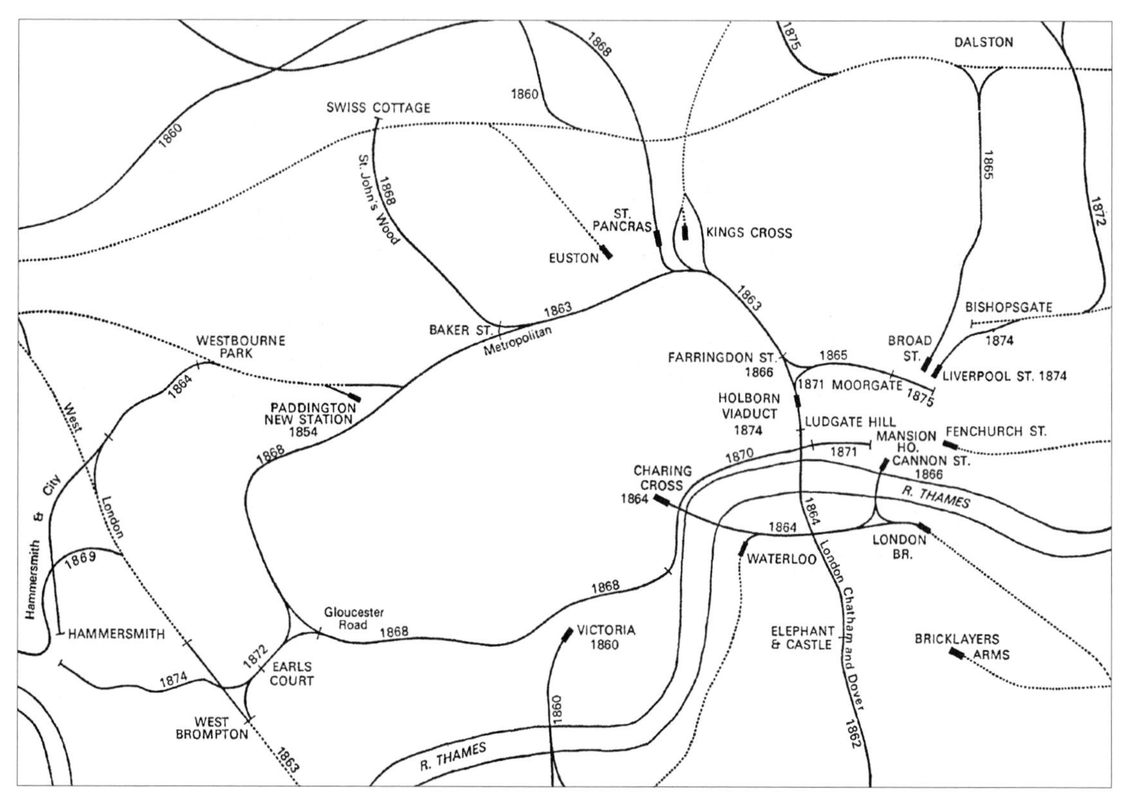

Railways in Central London opened between 1852 and 1875

Table 4: London railway termini

Terminus	Railway company	Date open
London Bridge	London & Greenwich	1836
Euston	London & Birmingham	1837
Fenchurch Street	London & Blackwall	1841
Waterloo	London & South Western	1848
King's Cross	Great Northern	1852
Paddington	Great Western	1854
Victoria	Victorian Station & Pimlico	1860
Charing Cross	South Eastern	1864
Cannon Street	London, Chatham & Dover	1866
St. Pancras	Midland	1868
Liverpool Street	Great Eastern	1874
Marylebone	Great Central	1899

between Wormwood Scrubbs and Uxbridge. They laid down an experimental track with a 9 in. diameter iron tube on a gradient of 1 in 120. The line was worked for nearly two and a half years, and during the first year the public were admitted free on two days a week. Among many engineers who visited the site were Robert Stephenson, I. K. Brunel and C. B. Vignoles. The London & Croydon Railway Company was formed under an Act of 1835 (5 & 6 Will.4 c.10) and the original locomotive-operated line was opened in 1839. In 1846 the company built an atmospheric railway alongside the original line, which was extended to Epsom in 1847. An interesting survival is the West Croydon engine house of the atmospheric railway bearing the date 1851, which was relocated as part of the Croydon Waterworks in Church Road and Surrey Street. The London & Croydon was one of only three atmospheric lines built, the others being Vignoles' Dublin & Kingstown Railway in Ireland and Brunel's South Devon Railway.

Railway civil engineering heritage in London consists principally of the main-line termini, Thames railway bridges, extensive brick viaducts, cuttings, embankments and tunnels. The historical engineering works are arranged chronologically by date of opening of the company's London terminus.

SIMMONS, J. *The Railway in Town and Country: 1830–1914*. David & Charles, Newton Abbot, 1986.

BAYLISS, D. A. *Retracing the First Public Railway*. Living History Local Guide No. 4. Living History, East Grinstead, 1985.

BARKER, T. C. and ROBBINS, M. *A History of London Transport*. Vol. 1, *The Nineteenth Century*, 1963; vol. 2, *The Twentieth Century to 1970*. Allen & Unwin, London, 1974.

JACKSON, A. A. *London's Termini*. David & Charles, Newton Abbot, 1969.

London's Overground Railways

1. London & Greenwich Railway

HEW 1725
TQ 380 773 to
TQ 328 802

This was the first steam railway to arrive in London. The company was formed under an Act of 1833 (3 & 4 Will.4 c.46), and the first section was completed in 1836 and the whole line in 1838. The Engineer to the Company was George Landmann, a retired Colonel in the Royal Engineers, and the contractor for the whole works was Hugh McIntosh. The railway route follows a straight line from London Bridge to Deptford and then takes a gentle curve to the Greenwich terminus. Two water-courses had to be crossed, namely Deptford Creek of the River Ravensbourne and the Grand Surrey Canal. The railway traverses the low-lying ground of Bermondsey and was built essentially as a viaduct line requiring considerable demolition of property. The letting of arches under the viaduct was of commercial benefit to the railway company. The railway was opened in three stages: Deptford to *Spa Road* in February 1836; Spa Road to London Bridge in December of that year; and Deptford to Greenwich on Christmas Eve 1838. The delay in getting

Greenwich Station

DENIS SMITH

into the Greenwich terminus was due to problems encountered in crossing Deptford Creek. Ten years later the railway began its eastward extension beyond Greenwich.

The brick viaduct carrying the railway is 3¾ miles long, and when built made a major impact on the then rural landscape. Tenders were invited in October 1833 and the contract was let to Hugh McIntosh. The first brick was laid on 4 April 1834, and within a few weeks some 400 men were laying 100 000 bricks a day. The completed viaduct comprised 878 arches, mostly of 18 ft span and 28 ft wide and 22 ft high. Within the viaduct itself the arches were mainly semi-circular, giving the viaduct the appearance of a Roman aqueduct. Where the railway crosses roads or watercourses different design solutions were adopted. An interesting survival is the railway skew bridge crossing *Abbey Street*, Bermondsey (formerly *Neckinger Road*), where the original fluted cast-iron Doric columns are still in place. The column capitals are spanned by a stone string-course providing the springing for the brick arch. Within the spandrel space four vertical brick walls support the four lines of rails. The voids between these walls are filled with concrete.

The volume of river traffic on the Ravensbourne necessitated the construction of a moveable railway bridge. The original railway bridge here (TQ 372 773) was of the drawbridge type, where the 26 ft span was raised by chains and a crab winch by eight men. It was completed in November 1838. The bridge of 1884 required 12 men to open it and it took an hour. The present bridge was installed in December 1963 and is an electrically operated vertical lift structure, the central platform weighing 40 tons. It was designed by A. H. Cantrell, Chief Civil Engineer of British Rail Southern Region, and was built by Sir William Arrol & Co., Glasgow.

THOMAS, R. H. G. *London's First Railway: The London & Greenwich*. Batsford, London, 1986.

BREES, S. C. *Railway Practice: A Collection of Working Plans and Practical Details of Construction*. John Williams, London, 1838.

London Bridge Station, Joiner Street Bridge

MALCOLM TUCKER

2. London Bridge Station and Joiner Street Bridge

HEW 2357
TQ 330 801

London Bridge Station was the first railway station in London. It was opened in 1836 as the terminus of the London & Greenwich Railway. It has been extended and rebuilt over the years as subsequent railway companies required accommodation at London Bridge. In June 1839 the London & Croydon Railway opened and ran into London Bridge Station over the London & Greenwich viaduct and tracks. In May 1842 the London & Brighton Railway (HEW 1573) used London Bridge over additional tracks. From 1847 the station was divided between the South Eastern Railway and the London, Brighton & South Coast Railway, and part of the 1866 Brighton Company's train shed survives on the south side of the station. The widening of the tracks in 1893–94 on a new viaduct provides an impressive facade to *Tooley Street*.

Joiner Street Bridge (HEW 2358) carries the extension lines to Cannon Street and Charing Cross Stations over *Joiner Street*, which enters the north side of London Bridge Station from *Tooley Street*. The deck of this bridge is carried on an interesting and early form of Warren girders (patented in 1848). The six parallel girders, 11 ft 6 in. apart, are of composite construction, combining inverted cast-iron equilateral triangles with rectangular wrought-iron tension members connecting the apexes in the lower chord. The spans vary from approximately 40 ft to 57 ft 3 in. and the depth of the girders is 4 ft 6 in. The bridge came to wider notice when it collapsed on 19 October 1850. Sir John Rennie and I. K. Brunel were asked to report on the failure and they decided against rebuilding to the same design. P. W. Barlow, the engineer to the railway, decided otherwise and the bridge remains to this day as a very rare survival from the evolutionary period of metal girder design.

The bridge failure at the south-eastern station, London Bridge. *Civ. Engr Architect. J.*, 1850, **13**, 390–91, 399.

The Builder, 1850, **8**, 524 (2 November), 553 (23 November), 574 (30 November).

3. London & Birmingham Railway

HEW 1092

The railway received its Act in 1833 (3 & 4 Will.4 c.36) for a railway with a terminus at Camden Station. Subsequently a site became available near Euston Square, and the company obtained an Act in July 1835 to extend the railway to the New Road. The first section of railway was opened from Euston to Boxmoor on 20 July 1837, and in October of that year it was opened as far as Tring. The whole line from London to Birmingham was opened on 17 September 1838. Robert Stephenson was the Engineer. The first Euston Station was London's first main-line terminus and was opened on 20 July 1837. There were just two platforms (arrival and departure), between which were two tracks for carriage standing. The twin pitched iron roof was carried on cast-iron columns with longitudinal iron arch arcading between the columns. Fox Henderson & Co. supplied the ironwork. A massive Doric portico, flanked by two small pavilions and two outlying lodges, was designed by Philip Hardwick. The portico, which became known as the Euston 'Arch', was used as the carriage entrance to the station. Its fluted Doric columns were of grey granite, 8 ft 6 in. diameter; each had a hollow core and one had a spiral staircase giving access to an archive store. The removal of the portico during redevelopment of the station in 1961–62 focused attention on, and raised widespread interest in, engineering conservation issues. Two lodges, designed by J. B. Stansley, the London & North Western Railway's architect, and dating from 1869, are still standing adjacent to the Euston Road.

JENKINSON, D. *The London & Birmingham: A Railway of Consequence.* Capital Transport, Harrow Weald, 1988.

4. Primrose Hill Cutting and Tunnel

HEW 2266

The Primrose Hill Cutting (TQ 292 830 to TQ 280 843) passes through the London Clay at an average depth of 20 ft below the surface and required massive, curved batter, retaining walls built in brick and bedded and backed in good concrete. Nevertheless, shortly after completion, the top of the western wall was pushed forward (in some places by more than 12 in.) by the swollen clay behind it. Temporary shoring and drainage holes were introduced. Robert Stephenson and Robert Dockray (the Resident Engineer) promptly introduced three remedial measures: cast-iron struts between the tops of the eastern and western walls at 20 ft centres; strong timbers embedded between the toes of the walls; and excavation behind the western wall to form brick vaults to reduce the pressure on the wall. The cruciform cast-iron struts, 96 in total, were supplied by Bramah and Cochrane of the Woodside Ironworks, Birmingham. The whole work was undertaken without any delay to the rail traffic. The struts were subsequently removed, although traces of them can still be seen in the retaining wall pilasters. The construction of this huge cutting, and its effect on Camden Town, was immortalised by the drawings of J. C. Bourne and by Dickens' graphic description in *Dombey and Son.*

With Primrose Hill Tunnel (TQ 276 843 to TQ 266 841) Stephenson again faced problems with the London Clay. The contractors, Jackson and Sheddon, began work in June 1834 by sinking four brick shafts of 8 ft internal diameter on the centre line of the tunnel. The contract was let for £120 000 for a tunnel just over half a mile long. The tunnel has a circular brick arch of 23 ft 9 in. span, curved side walls resting on stone skew backs, and a brick invert arch. The height from the rails to the crown of the arch is 21 ft 6 in. By November 1834 the contractors were defeated by clay pressure problems, and Robert Stephenson took over direct control of the work, which was completed at a cost of £280 000—over twice the original contract price. The tunnel has Italianate portals, which Simms suggests were 'expressive of the strength which Railway works should, it is presumed, be characterized'.

DEMPSEY, G. D. Description of the mode adopted for repairing and supporting the western retaining wall of the L & B Railway. *R. Engrs Prof. Papers*, 1844, Quarto, **7**, 160–64.

SIMMS, F. W. *The Public Works of Great Britain*. John Weale, London, 1838, 22.

5. Winding Engine Vaults, Camden Town

HEW 2267
TQ 289 839

Even with the Primrose Hill Cutting the gradient (averaging 1 in 85) facing trains on leaving Euston Station was too great for the locomotives of the day. Steam-powered winding engines with rope haulage were used to draw trains out of Euston and were placed underground in a great barrel-vaulted brick chamber, with two impressive chimneys rising 132 ft 4 in. above rail level, close to the Regent's Canal. Two 60 hp engines, boilers and the winding machinery were supplied by Maudslay's of Westminster Bridge Road. The rope was 3744 yd long, of 7 in. circumference and weight 11.5 tons, and to keep it taut it passed round a pulley on a moveable counterweighted carriage before emerging on the surface between the rails. From October 1837 trains from Euston were drawn up the inclined plane to Camden Station, where locomotives waited to take the trains onwards. But the rope haulage system was short lived, and as locomotives improved the system was abandoned in April 1844. Soon afterwards the engines were sold to power a flax mill in Russia. The magnificent brick barrel-vaulted chambers survive and were pumped dry and repaired in 1998. They are Listed Grade II, and a variety of schemes for their adaptive re-use has been discussed, but as they lie under the main railway line the options are limited.

DOCKRAY, R. B. Description of the Camden Station of the London & North-Western Railway. *Min. Proc. Instn Civ. Engrs*, 1849, **8**, 164–85.

SIMMS, F. W. *The Public Works of Great Britain*. John Weale, London, 1838, 2.

6. Round House, Camden

HEW 238
TQ 283 843

The London & Birmingham Railway Company decided to build two locomotive sheds on the Camden site: a rectangular structure for passenger locomotives, and a circular one for goods engines. The Round House, as it became known, was built 1846–47, is 160 ft in diameter and provided 24 lines of railway radiating from the

The Round House, Camden

WENDIE TEPPETT

centre, each long enough to accommodate a locomotive and its tender. The edge of the turntable had a cast-iron curb with slots in it only at the ends of the rails on the turntable so that a locomotive could only enter the turntable when these slots were aligned with the outer rails. The substructure of the engine house is of brick vaulting up to the underside of the rails. The top of the conical roof is 67 ft above the rails and light is admitted through a 30 ft diameter central light and a 10 ft wide glazed section placed halfway down the length of the rafters. The ends of the principal rafters are carried on a cast-iron curb. This building had an associated coking shed, office, stores and fitters' shop. The contractors were Branson & Gwyther of Birmingham. The Round House is Listed Grade II* and has had several uses in the arts field since the railway ceased using it for its original purpose.

DOCKRAY, R. B. Description of the Camden Station of the London & North-Western Railway. *Min. Proc. Instn Civ. Engrs*, 1849, **8**, 165–85.

7. London & Blackwall Railway

HEW 2271
TQ 334 809 to
TQ 388 802

The company was formed as *The Commercial Railway (London & Blackwall)* under an Act of 1836 (6 &7 Will.4 c.123). The proposed railway threatened demolition of the homes of 2850 people. The Engineers were George Parker Bidder and Robert Stephenson. In December 1837 Bidder recommended rope-haulage traction, partly because of the Euston incline experience and partly for environmental reasons. The railway was carried on a brick viaduct and was 3½ miles long. The foundations were deficient and several accidents ensued. The railway was opened on 4 July 1840 and the first London terminus was at the *Minories* and the Blackwall terminus was at Brunswick Wharf (with which Bidder had been involved). In 1841 the London terminus was moved to *Fenchurch Street* (HEW 441). The line was laid to a gauge of 5 ft. The steam winding-engines at the London terminus were built by Maudslay, Sons and Field. The line had seven stations served by four trains per hour each way.

The extensive brick viaducts and bridges have now been renovated and adapted for the Docklands Light Railway. The Limehouse Viaduct, built in 1838–40, comprises three large arches, each of 87 ft span, crossing the

WENDIE TEPPETT

London & Blackwall Railway Viaduct

Regent's Canal as it enters the Limehouse Basin. The line was converted to locomotive traction in 1848, and for a period the viaduct was covered by an iron roof to reduce the fire risk to wooden ships in the dock below. The disused Limehouse Station in *Three Colt Street* is the only surviving example of an original 1840 London & Blackwall Railway station. The platforms were removed in 1926, when the station closed, but much of the ground-level building remains.

8. Waterloo Station and International Terminal

HEW 2270
TQ 311 798

Waterloo Station was opened in 1848 by the London & South Western Railway. It replaced the company's earlier London terminus at Nine Elms. Initially housing only three platforms, the station was enlarged by extensions in 1862, 1878 and 1885. During 1900–22 a major rebuilding programme took the number of platforms up to 21, making the station the largest in the United Kingdom. Of the nineteenth century station only the roof over platforms 16–21, dating from 1885, remained. Platforms 1–15 and a spacious new concourse were covered by

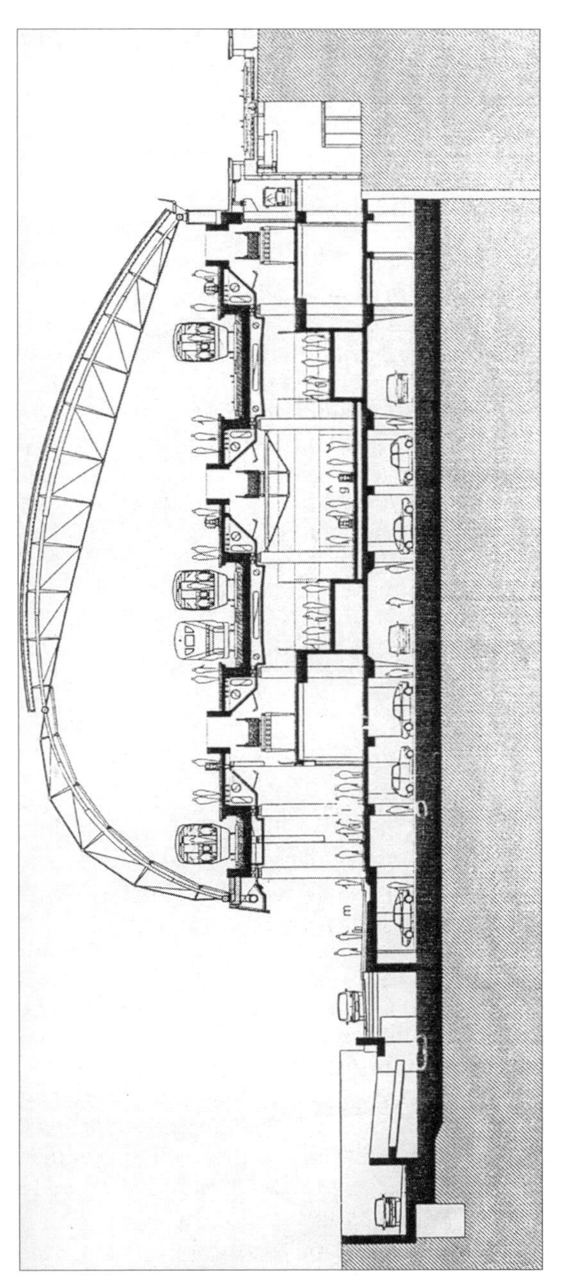

Waterloo International Terminal, circulation area

560 000 sq. ft of ridge-and-furrow form, with steel truss girders on steel columns.

The rebuilding included new offices for the railway company and the imposing Victory Arch over the main entrance commemorating the London & South Western Railway's employees who had lost their lives in the First World War. The cost of reconstruction was just over £2¼ million.

By the mid-1980s a scheme to site the London terminus for passenger services to Europe via the Channel Tunnel at Waterloo was well developed, and powers to build a new station were included in the 1987 Channel Tunnel Act. The scheme involved the removal of the last remaining section of the nineteenth century train shed roof over platforms 16–21 and of the Armstrong car hoist serving the Waterloo & City Railway. In the domestic station, four new platforms were built to compensate for the loss of those that fell within the footprint of the International Terminal. The resulting layout consists of 19 platforms for domestic and five for international services. The international terminal includes a basement car park. Above this is a complex of concourse and service areas and, at the top level, five platforms.

The train shed is formed as a series of asymmetrical three-pinned arches, each consisting of two steel bowstring trusses. It is curved and tapered in plan, the maximum span being 48.5 m at the buffer stop end, reducing to 32.7 m at the far end. The asymmetry was introduced to provide adequate structural clearance over the outermost track, alongside platform 24, the narrow and more steeply inclined part of the roof over this platform being glazed from springing level to the crown. Stainless steel cladding, with intermittent glazed bays, covers that part of the roof carried on the longer trusses. The overall length of the shed is approximately 400 m.

Following preliminary works during the year the main construction began in December 1990 and was completed in May 1993 at a cost within the budget of £143 million at 1989 prices.

FRANKLIN, J. L. Waterloo International Terminal. *Proc. ICE, Transport*, 1999, **135**, 55–91.

HUNT, A. J., JONES, A. C., OTLET, M. and DEXTER, D. I. Waterloo International trainshed roof. *Struct. Engnr*, 1994, **72**, No. 8, 123–29.

FAULKNER, J. N. and WILLIAMS, R. A. *The LSWR in the Twentieth Century*. David & Charles, Newton Abbot, 1988.

JACKSON, A. A. *London's Termini*. David & Charles, Newton Abbot, 1985, 210–42, 361–63.

BIDDLE, G. and NOCK, O. S. *The Railway Heritage of Great Britain 1983*. Michael Joseph, London, 1983, 184.

WILLIAMS, R. A. *The London & South Western Railway*. David & Charles, Newton Abbot, vol. 1, 1968; vol. 2, 1973.

CAIRNS, J. F. The largest railway terminus in Great Britain. *Railway Mag.*, 1922, **50**, 298–313.

DAVEY, P. Waterloo International. *Architect. Rev.*, 1993, **193**, No. 1159, 26–44.

9. Great Northern Railway

HEW 1921

William Cubitt (1785–1861) was appointed Engineer to this project in September 1844 and was assisted by his son Joseph. The company received its Act in 1846 (9 & 10 Vict. c.71) and sought to bring to London agricultural produce from Bedfordshire and, later, the Fens, and coal from Yorkshire. The Yorkshire coal traffic was a mainstay of the Great Northern Railway. Until 1850 nearly all of London's coal had come by sea from Northumberland and Durham, but within five years 30% was coming by rail into King's Cross to be redistributed by road and the Regent's Canal. The buildings in the goods station reflect this traffic. In August 1850 they opened a goods station and a temporary passenger station on the north bank of the Regent's Canal in order to capture some of the traffic created by the Great Exhibition of 1851. They brought the railway under the canal by the *Gasworks Tunnel*. King's Cross Station was opened in October 1852.

10. King's Cross Station

HEW 302
TQ 305 831

The completion of this station, built in 1851–52 to the design of Lewis Cubitt, was delayed by the need to demolish the fever hospital on the site. Its functionalist lines have been much admired by twentieth-century architectural critics. The station platforms are spanned by two series of arches, each 105 ft wide, delivering their load onto a central longitudinal brick wall with arched openings. The roof arches are at 20 ft centres and the train shed is 800 ft long. The original arches were of laminated timber construction having a cross-section 10½ in. wide and 24 in.

deep. The eastern roof, which was not buttressed by an office building, suffered distortion, and these arches were replaced with wrought-iron arches in 1869–87, replacement of the western roof following in 1887. The south end of the station is flanked by a simple yellow stock brick facade, which reflects the arches of the train shed. The clock in the central tower had previously been exhibited in the Crystal Palace exhibition. The station is Listed Grade I.

HUNTER, M. and THORNE, R. *Change at King's Cross*. Historical Publications, London, 1990.

11. Gasworks Tunnel under the Regent's Canal

HEW 365
TQ 305 835

North of King's Cross station the main line dips at 1 in 100 to pass under the Regent's Canal and then climbs at a similar gradient, mostly within the *Gasworks Tunnel*, which comprises three parallel bores, each 530 yd long. The central bore of 1852 was supplemented in 1874–78 on the east side and in 1889 on the west side to relieve the congestion of 'The Throat'. Since 1977, modern signalling has made possible the abandonment of the eastern bore. The tunnels are of conventional vaulted brickwork, but beneath the canal the tunnel roofs are of shallow cast-iron beams to minimise the constructional depth. On the central tunnel of 1852 the cast-iron beams, or plates, are each 3 ft wide with three upstand parabolic ribs with a maximum depth of 1 ft 3 in. at mid-span. The overall length of each of the plates under the canal is 31 ft, giving a clear span of 25 ft. The plates, 1½ in. thick, are bolted together with joints caulked by iron cement and all covered with 2 ft of puddled clay above the tunnel soffit. The later tunnels of 1878 and 1892 are basically similar to the original, but the cast-iron plates are more substantial and each weighs 8 tons, compared with 4¼ tons in the 1852 structure. The engineer for the first tunnel was Joseph Cubitt.

12. The Granary, King's Cross Goods Yard

HEW 674
TQ 301 836

This fascinating group of buildings represents an early example of the integration of rail, road and canal traffic.

The principal building is the *Goods Shed* of 1850 with the magnificent *Granary* of 1851–52 at the south end. The Granary facade is of stock brick with stone dressings, six storeys high, with timber floors and roofs on cast-iron columns and beams. The architect was Lewis Cubitt and the building contractor was John Jay. In front of the granary a *Basin* (1850) connected with the Regent's Canal. From the basin, four channels enabled canal boats to pass under both the Granary and the Goods Shed railway platforms, allowing goods to be transhipped vertically from rail to canal. Similarly, the *Eastern Coal Drops* (1851) and the *Western Coal Drops* (1856) allowed rail coal wagons at high level to discharge coal into horse-drawn road vehicles below. Also on site are the *Fish and Coal Offices* and the *Midlands Goods Shed*, an 1850 carriage shed which was altered in the 1880s.

13. Dollis Road Railway Viaduct

HEW 2352
TQ 246 911

The Edgware Highgate & London Railway was authorised by an Act of 1862. John Fowler and Walter Brydone (the Great Northern Railway's Chief Engineer from 1855 to 1861) were the engineers for the line, which was transferred to the Great Northern Railway in July 1867. Smith, Knight & Co. were the contractors. About 9 miles long, the railway ran from Finsbury Park to Edgware, via Finchley and Mill Hill. It opened on 22 August 1867.

The brick viaduct on which the line crosses *Dollis Road* and the adjacent *Dollis Brook* has 13 segmental arches, each of about 32 ft span at springing level, carried on tapered piers. Each pier is pierced by an opening with an arched soffit and dished at its invert. The track is carried at about 60 ft above water level.

Although built for double track, the viaduct initially carried only one line. As part of the 1935–40 works programme, sponsored by the Government, the route was doubled and electrified by the London & North Eastern Railway, successor to the Great Northern Railway. London Transport trains worked over one of the tracks from 18 May 1941; the second track was lifted soon afterwards.

In the early 1950s a scheme to extend the line to Bushey Heath in Hertfordshire, on which work had started just before the Second World War, was abandoned. The track

Viaduct over Dollis Brook at Mill Hill East

MALCOM TUCKER

beyond Mill Hill East was lifted in 1964, leaving that station as a terminus of London Underground's Northern Line.

CROOME, D. F. and JACKSON, A. A. *Rails Through the Clay*. Capital Transport, Harrow Weald, 1993, 2nd edn.

JACKSON, A. A. *London's Local Railways*. David & Charles, Newton Abbot, 1978.

CARTER, E. *An Historical Geography of the Railways of the British Isles*. Cassell, London, 1959.

HOPWOOD, H. L. The Edgware, Highgate & London Railway. *Rail. Mag.*, 1919, **45**, 1–7.

14. Paddington Station

HEW 301
TQ 265 814

The Great Western Railway Company received its Act in 1835 (5 & 6 Will.4 c.107); the first section was completed in 1838 and reached Bristol in 1841 (see also Chapter 9). Its London terminus is Paddington Station.

The company's first London terminus was a temporary structure opened on 4 June 1838. It was sited just to the north-west of the present station, which was brought into service, although incomplete, on 16 January 1854. The roof then comprised three semi-elliptical spans of wrought-iron arch ribs, supported on cast-iron columns carrying a light and airy roof of glass and corrugated iron. From the south side the spans are 69 ft 7 in., 102 ft 6 in. and 68 ft; the covered length of 700 ft includes two 'transepts', each 50 ft wide.

Isambard Kingdom Brunel, the Engineer for the Great Western Railway since the company's inception, engaged the architect Matthew Digby Wyatt to detail the ornamentation. Fox Henderson and Co. were the contractors for the train shed. Construction continued for several years after opening, and the total expenditure on the whole station, including the goods station and engine house, reached £650 000 by 1862.

A fourth span in steel was added to the train shed in 1915–16. At the same time a start was made to replace Brunel's original cast-iron columns with steel columns fabricated to a similar design. This work was interrupted by the First World War and was not completed until 1924.

Successive replacements of the roof glazing and cladding were made during the second half of the twentieth century, using modern materials sympathetic to the original design. During the 1990s repairs have included conservation and strengthening work to the structural framework.

The Great Western Royal Hotel, which forms the *Praed Street* frontage of Paddington Station, was designed by Philip Hardwick and was opened in June 1854.

CONNELL, G. S. The restoration of Brunel's Paddington Station roof, *Proc. Instn Civ. Engrs*, 1993, **93**, 10–18.

THORNE, R. Paddington Station. *Architects J.*, November 1985, 44–58.

JACKSON, A. A. *London's Termini*. David & Charles, Newton Abbot, 1985, 2nd edn.

MACDERMOT, E. T. *History of the Great Western Railway*. Ian Allan, London, 1964.

15. Wharncliffe Viaduct, Hanwell

HEW 591
TQ 150 804

The River Brent was the first major obstacle facing Brunel's westward route from Paddington. The viaduct, carrying the Great Western Railway main line 65 ft over the Brent valley, was Brunel's first major structural design and was also the first contract to be let on the Great Western Railway. The contractors for this major project were Grissell & Peto, who began work in February 1836. The elegant brick structure is 900 ft long and comprises eight arches, each of 70 ft span and 17 ft 6 in. rise. Hollow brick piers taper up to projecting stone cornices, which were used to support the arch centring. The piers of the original structure are 30 ft overall at ground level, providing a deck capable of taking two lines of broad-gauge track. Lord Wharncliffe's Coat of Arms decorates the central pier on the south side. Wharncliffe was Chairman of the Great Western Railway and for this reason the structure is often known as the Wharncliffe Viaduct. On the night of 21 May 1837, when construction was nearly complete, a serious slip in the clay of the embankment occurred on the north side. The slip was 50 ft. wide and

Wharncliffe Viaduct, Hanwell

J. B. POWELL

increased slowly over four months. Brunel took appropriate action, and the viaduct was completed in the summer of 1837 at a cost of £355 000. In 1874 the structure was widened on the north side from 33 ft to 55 ft by adding an additional row of brick piers and arches. The viaduct is a Listed Grade I structure.

PUGSLEY, SIR A. G. *The Works of Isambard Kingdom Brunel*. Institution of Civil Engineers/University of Bristol, London/Bristol, 1976.

MACDERMOT, E. T. *History of the Great Western Railway*. Ian Allan, London, 1964.

16. Three Bridges, Hanwell

HEW 2269
TQ 143 796

This unusual three-level structure comprises a road, *Windmill Lane*, at the top level, the Grand Union Canal at the middle level, and a branch of the Great Western Railway at the lower level. Originally only the road was there but in 1800–01 the Grand Junction Canal Company arrived, with the canal on an embankment, and built a bridge to carry the road over the canal. By 1841 the Great Western Railway line from Paddington to Bristol was completed and the railway naturally wished to acquire a link with the canal, to provide a link with the Thames. The railway and canal met at Bulls Bridge, near Hayes, and a transhipment depot was built there, although the 6-mile journey to the Thames was slow as 11 locks were involved. By 1850 the tonnage of coal transferred from rail to barge had reached 50 000 tons, and a rail connection with Brentford was suggested. In 1855 an Act was obtained for a railway and a new dock at Brentford and the Great Western & Brentford Railway was formed. The line was the last railway project of I. K. Brunel. Construction work began in early 1856 and the railway opened for goods traffic on 18 July 1859 and for passengers one year later. The line, a little over 3 miles long, was designed as a double broad-gauge track, but only one track was laid in 1859. In October 1861 an additional mixed-gauge track was laid, and in 1875 both tracks were reduced to standard gauge. Both the canal and railway Acts had required their alignments to avoid Osterley Park. The canal and railway therefore occupied the same narrow site between the Park and St. Bernard's Hospital. This led to the unique

Three Bridges, Hanwell

DENIS SMITH

structure we see today. The railway and canal cross at an angle of about 35°, and the canal and road at right angles. The cast-iron aqueduct consists of two spans with an overall length of 59 ft. The trough, 18 ft wide with an external depth of about 7 ft, is carried above the railway on brick piers. A towpath lies on the north bank of the canal, with 10 ft headroom to the soffit of the road bridge. The road bridge comprises cast- iron girders carrying a recent concrete deck. Although I. K. Brunel was Engineer to the Great Western & Brentford Railway, his stressful occupation with the launch of the *SS Great Eastern*, his absence on sick leave in Europe at this time, together with the opening of the Royal Albert Bridge at Saltash in May 1859, means that others in his office must be given credit for most of the work on the Great Western & Brentford Railway. Brunel died on 15 September 1859.

PERRET, D. Three Bridges, Hanwell: an outline history. *London's Ind. Archaeol.*, 1994, **GLIAS 5**, 35–38.

17. Victoria Station

HEW 455
TQ 288 789

The London, Brighton & South Coast Railway's first London terminus was designated the Pimlico Terminus, although sited in Battersea New Town on the south side of the Thames. It was a temporary structure and was opened in March 1858. In that year the Victoria Station & Pimlico Railway, with support from the London, Brighton & South Coast Railway (including a major share of the capital), obtained an Act for a bridge over the river and a line to a new station, Victoria, on the site of the Grosvenor Canal basin between *Buckingham Palace Road* and *Wilton Road*. The London, Brighton & South Coast Railway terminus, on the western side, came into use on 10 October 1860. The train shed consisted of a ridge-and-furrow type roof carried on wrought-iron lattice girders. There were ten tracks and six platform faces. The Grosvenor Hotel, an independent enterprise, opened in the following year.

The eastern side of Victoria Station was originally built as a separate station, shared by the London Chatham & Dover (HEW 1760) and the Great Western Railways. It opened in 1862. By 1869, following the widening of Grosvenor Bridge (originally named Victoria Bridge), there were nine platform faces served by nine tracks, of which four were laid as mixed gauge. This part of the station is the most interesting to the historian of structural engineering, as John Fowler's original roof, comprising two elegant lattice arch spans of 124 ft and 117 ft, 740 ft long, remains over the platforms now numbered 1 to 8.

During 1901–08 the London, Brighton & South Coast Railway's station was rebuilt and extended by lengthening beyond Eccleston Bridge, the expansion being constrained laterally by the presence of the hotel on one side and the London, Chatham & Dover Railway station on the other. A new lattice girder roof replaced the original, and a new station frontage, including a hotel block designed by the London, Brighton & South Coast Railway's Chief Engineer, Sir Charles Morgan, to blend with the Grosvenor, was built.

Meanwhile, in 1899 the London Chatham & Dover Company had merged with the South Eastern to become the South Eastern & Chatham Railway. Not to be outshone, the South Eastern & Chatham Railway built a new station frontage, designed by A. W. Blomfield and opened in 1909. The Engineer for the first London, Brighton & South Coast Railway station was Robert Jacomb Hood and the architect for the Grosvenor Hotel was J. T. Knowles.

JACKSON, A. A. *London's Termini*. David & Charles, Newton Abbot, 1985, 267–302.

TURNER, J. T. H. *The London Brighton & South Coast Railway*. Batsford, 1977–79, 3 vols.

18. Midland Railway

HEW 1924

This railway was formed by the amalgamation in 1844 of three existing companies, namely *The Midland Counties*, *The North Midland and Birmingham* and *The Derby Railway*, giving the new company about 200 miles of railway. W. H. Barlow was consulting engineer to the Midland Railway Company. Access to London was initially solely dependent on the London & North Western Railway into Euston, but in 1857 the Midland Railway opened the Leicester–Hitchin railway, giving them a second route over the Great Northern Railway line into King's Cross. In 1862–65 the company established a separate goods station at Agar Town, north of the Regent's Canal, and its own main line extension from Bedford to St. Pancras was constructed in 1863–68. Like the Great Northern Railway, it had a large traffic in coal from the East Midlands, with surviving coal drops of the company's own distinctive design along Pancras Road and a coal basin on the canal. Somers Town Goods Station was added alongside the St. Pancras passenger station in 1887. On two levels, with hydraulic wagon hoists, it included an extensive Potato Market. A similar two-level Coal Depot was added to the north in 1896. These massively constructed structures in cast iron, steel and engineering brick have mostly been demolished for the British Library and for a housing estate, respectively, but parts survive, particularly the red brick Gothic perimeter walls. Nearby are some plate girder bridges of 1867, by the Butterley Company.

19. St. Pancras Station Train Shed

HEW 237
TQ 301 830

When opened on 1 October 1868, the Midland Railway's London terminus Train Shed, now a Listed Grade I building, was the world's largest clear span roof. Designed by W. H. Barlow with the assistance of R. M. Ordish, the roof is formed by a series of wrought-iron tied arch ribs having a clear span of 240 ft. The unprecedented span led Barlow to make a 13 ft 4 in. span scale-model load test of the arch. The ribs, of lattice form, are spaced at 29 ft 4 in. centres and the overall length of the shed is 675 ft. There is a glazed, pitched, extension of 15 ft over the concourse, linking the Train Shed to the

St. Pancras Station Train Shed

WENDIE TEPPETT

hotel. The arch tie girders serve as floor beams for the station and are supported on a grid of 720 cast-iron columns carried down to street level. The need to cross the Regent's Canal at a suitable height determined the track level of 17 ft above the Euston Road. The space below the platforms was used for storing the Burton beer traffic, the modular grid of the columns being chosen to optimise the storage of the barrels. Waring Brothers were the main civil engineering contractors. The ironwork was supplied by the Butterley Company, who assembled the ribs, each weighing 55 tons, using a massive timber centring framework travelling on rails. The tender sum for the roof was £116 720.

The Train Shed is set behind the exuberant facade of the Grand Midland Hotel, opened between 1873 and 1876. The hotel was converted to offices in 1935. The site is about to be developed as part of the Channel Tunnel Rail Link.

SIMMONS, J. *St. Pancras Station.* Allen & Unwin, London, 1968.

BARLOW, W. H. Description of the St. Pancras Station and roof, Midland Railway. *Proc. Instn Civ. Engrs*, 1870, **9**, 78–93.

20. Liverpool Street Station

HEW 303
TQ 339 804

The Eastern Counties Railway was authorised by Act of Parliament in 1836 to construct a railway between London and Norwich and Yarmouth via Ipswich. The first section opened from Mile End to Romford on 20 June 1839, and was later extended to Shoreditch where a terminus was opened on 1 July 1840. Between 1840 and 1845 the Northern & Eastern Railway built a line from Stratford to Cambridge, which also terminated at Shoreditch. This station had single departure and arrival platforms, each 256 ft long. Between 1848 and 1849 Sancton Wood rebuilt the station and the terminus was renamed Bishopsgate. When the new Liverpool Street Station was built in 1874 the old Bishopsgate terminus went out of use, and in 1881 became a goods station, which was destroyed by fire in 1964. As early as 1845 the Eastern Counties Railway had proposed an alternative terminus nearer to the City, but it was not until after 1862 when the leading railway companies in East Anglia had amalgamated into the Great Eastern Railway Company that a scheme was agreed and sanctioned by Parliament in 1864—a low-level station in Liverpool Street to the east of the North London Railway terminus then under construction.

Ten years elapsed before the new project was realised. The new station had nine platforms, subsequently increased to ten. The new terminus was opened to local traffic on 2 October 1874 and to all traffic on 1 November 1875. Alongside the main arrival platform were built the main booking hall, and refreshment and waiting rooms, and above them the general offices of the Great Eastern Railway Company. The engineer was Edward Wilson (1820–77), Consulting Engineer to the Company, the contractors were Lucas Brothers, and the Fairbairn Engineering Company of Manchester supplied the ironwork.

From 1 February 1874 the Metropolitan Railway used platforms 1 and 2 of the Great Eastern station until the Metropolitan Railway opened their own station on 12 July 1874, which was called Bishopsgate until 1 January 1909. In 1879 the Great Eastern Board appointed C. E. Barry as Architect and Works Surveyor of the hotel to be built, and which was opened on 26 May 1884. Under an Act of 1888 work began in 1890 to construct eight extra platforms on the east side of the station, with its separate concourse and passenger facilities, which was opened on 2 April 1894. These works were carried out under John Wilson (nephew of Edward Wilson), Engineer-in-Chief to the Great Eastern Railway (appointed 1883), and the architect was W. N. Ashbee.

In 1912 the Central London Railway was extended to Liverpool Street and in 1927 the Post Office Railway, which was opened between Paddington and Whitechapel, was also provided with its own station at Liverpool Street. The station was extensively remodelled from 1983, with the passenger hall being opened out, new walkways installed and parts of the *Broadgate* scheme constructed over the tracks.

MOORCROFT, R. L. (Chief Architect British Railways Board). *Evidence given to the Planning Inquiry: Redevelopment of Liverpool Street and Broad Street Station*. November 1976.

Railway Magazine, December 1898, 516; February 1906, **18**, 89.

The Engineer, 11 June 1875, **39**; 12 October 1894, **78**.

21. Commercial Street Bridge, Shoreditch

HEW 70
TQ 335 822

Following the absorption of the Eastern Counties Railway into the Great Eastern Railway in 1862, the main line was to run into the new low-level terminus in Liverpool Street (see above). The approach is in cutting, over which *Commercial Street* is carried on this single-span cast-iron bridge. There are 11 cast-iron arched ribs forming a skew span of 80 ft, with a rise of 8 ft 2 in., supporting a deck made up partly of brick jack arches and partly of wrought-iron dished plates. The spandrels are filled with diagonal members as part of the monolithic casting. Construction was completed in time for the opening of the first stage of Liverpool Street Station on 2 February 1874.

PHEW site record form.

JACKSON, A. A. *London's Termini*. David & Charles, Newton Abbot, 1985.

London's Underground Railways

The problem of moving large numbers of people around in a heavily built-up and already rapidly expanding conurbation led to London becoming the location of the world's first underground urban railway—the Metropolitan, opened in 1863. Later, the development of electric traction and the availability of lifts to transport passengers vertically led to the construction of a network of deep-level railways, beginning with the City and South London Railway which opened in 1890, again a world first.

By 1907 what are now known as the Northern, Central, Bakerloo and Piccadilly Lines were complete in central London. During the next half century these and the Metropolitan and District Lines were extended by stages into the suburbs and the surrounding country, but no new underground line crossing the capital was built until the construction of the Victoria Line in the 1960s. The last 30 years of the twentieth century saw the building of the Jubilee Line (partly converted from a branch of the Bakerloo Line), its extension into the Docklands area and East London, and the extension of the Piccadilly Line to Heathrow Airport.

22. Metropolitan Railway

HEW 2273

The desirability of a link between the main line stations on the northern and western sides of London and the City had been apparent from the early 1850s. In 1854 Parliament passed an Act authorising the formation of the Metropolitan Railway Company and the building of the world's first underground railway, from Paddington to *Farringdon Street* via King's Cross. As far as possible the route was to follow existing roads so as to minimise demolition of buildings and compensation payments to their owners. Difficulty in raising capital delayed the start of construction until 1860. Except for a 728 yd tunnel beneath Mount Pleasant, at Clerkenwell, the section of line between King's Cross and Farringdon Street was in open cutting. From Paddington to King's Cross it was built by the cut-and-cover method, mainly beneath *Praed Street* and the *Marylebone* and *Euston Roads*. Where sufficient depth was available an elliptical brick arch was built spanning between the brick walls of the trench. Elsewhere the usual roof structure was of cast-iron girders supporting brick jack-arches. Shallow construction facilitated ventilation and passenger access by stairways; a disadvantage was the need to divert

Metropolitan Railway, Farringdon

WENDIE TEPPETT

water and gas services, sewers and telegraph lines. The Fleet Ditch sewer near *Farringdon Street* gave rise to the first serious accident during construction when it burst on 18 June 1862, flooding the line only a few weeks after the first trial run over the full length of the railway had taken place. John Fowler was the engineer for the line. The contractors were John Jay (Euston Square to the City) and Smith and Knight (Paddington to the west side of Euston Square).

Public services began on 10 January 1863. In the first few months condensing locomotives and rolling stock hired from the Great Western Railway were used, the Metropolitan's track having been laid as mixed gauge in order to allow the running of both broad- and standard-gauge trains. Following a disagreement between the Metropolitan and the Great Western, standard-gauge stock from the Great Northern Railway replaced the Great Western trains until the Metropolitan acquired sufficient locomotives and carriages of its own. The broad-gauge track was removed in 1869.

In 1860 Fowler had commissioned from Robert Stephenson and Company an experimental fireless locomotive in an attempt to avoid pollution from exhaust in the tunnels. The engine was a failure and the idea was abandoned. Unfortunately, the condensing locomotives were only partly successful in alleviating the nuisance, making it necessary for additional ventilation openings to be installed during the early years of the railway's operation. Poor conditions underground persisted until the introduction of electric traction in 1905, power being provided from the Metropolitan's own generating station at Neasden. By then the railway had been extended well into the country to the north-west of London. At its eastern end, within five years of opening the Metropolitan had developed links with the Great Northern, the Midland and the London, Chatham & Dover railways. Between King's Cross and Moorgate the line had been quadrupled in length. An extension to Liverpool Street was opened in 1875 and Aldgate was reached the following year.

Much rebuilding of stations on the original route of the Metropolitan has taken place. In 1984, renovation of the *Baker Street* Station tunnel, now used by Circle Line and

Hammersmith and City trains, was completed, resulting in a credible representation of its original appearance, albeit without the steam and smoke which would have pervaded the space in 1863. However, at Farringdon Station (HEW 440) (TQ 316 819) John Fowler's iron and glass overall train shed roof of 1865 survives.

JACKSON, A. A. *London's Metropolitan Railway*. David & Charles, Newton Abbot, 1986.

BARKER, T. C. and ROBBINS, M. *A History of London Transport*. Vol. 1, *The Nineteenth Century*, 1963; vol. 2, *The Twentieth Century to 1970*. Allen & Unwin, London, 1974.

23. Metropolitan District Railway

HEW 2274

In 1864 Acts authorising extensions of the Metropolitan Railway and the establishment of the Metropolitan District Railway received Royal Assent. An important objective was to complete what is now known as the Circle Line, but this was not achieved until 1884. By late 1868 an extension of the Metropolitan from Praed Street, Paddington, was opened and linked to the Metropolitan District at South Kensington, the Metropolitan District Railway having by then reached Westminster. Fowler was again the engineer, and the method of construction was principally cut and cover for the subsurface sections. Experience of the ventilation problems on the Metropolitan Railway led to relatively more of the Metropolitan District Railway being built in permanent open cut. Features of interest are the 421 yd tunnel under Campden Hill and a conduit carrying the Ranelagh sewer (formerly the River Westbourne) over the platforms at Sloane Square Station. The contractors for the work were Peto and Betts, Kelk and Waring Brothers. Extension eastward from Westminster had to be coordinated with the construction of the Victoria Embankment (see p. 28) and the line reached Mansion House in 1871. Difficulty in raising funds and bickering between the Metropolitan and the District Railways delayed completion of the Circle Line through Aldgate, which was finally opened on 6 October 1884.

Meanwhile, extension westwards had already begun in 1869, when a line via Earl's Court established a station adjacent to that of the West London Railway at West

Brompton. Branches from Earl's Court were completed to Richmond (1877), Ealing Broadway (1879) and Wimbledon (1889). John Wolfe Barry and Sir John Hawkshaw were engaged on the completion of the Circle Line.

BAKER, B. The Metropolitan and Metropolitan District Railways. *Min. Proc. Instn Civ. Engrs*, 1884–85, **81**, 1–74.

24. Earl's Court Station

HEW 2275
TQ 254 784

When the Metropolitan District Railway was first opened to West Brompton there was no station at Earl's Court and local residents petitioned for one. A small station was opened on 31 October 1871, but this was destroyed by fire on 30 December 1875. This provided the opportunity to build the present station on a site to the west of the old structure. Work began in 1876 and the station opened on 1 February 1878. The main feature of the present station is its impressive, and economic, overall roof in wrought iron designed by John Wolfe Barry. The pitched roof trusses are of 96 ft span, spaced at 20 ft centres. One end of each roof truss rests on a single steel roller. Each rafter has five purlins which are vertical lattice girders with a curved lower chord. The main tie rods are 2¼ in. diameter wrought-iron rods. The train shed, the *Earl's Court Road* frontage (architect H. W. Ford) and the later *Warwick Road* entrance are Listed Grade II.

WALMISLEY, A. T. *Iron Roofs*. E. & F. Spon, London, 1888, 16–17.

CONNOR, P. *Going Green*. Capital Transport, Harrow Weald, 1994.

25. London's Early Tube Railways

HEW 2276

London, especially to the north of the Thames, is extensively underlaid by a stiff blue clay, an ideal medium in which to construct shield-driven tunnels. Greathead shields were used to build most of the early tubes, the excavation being accomplished by hand mining. The tunnel linings were normally rings made up of cast-iron segments assembled *in situ*, bolted together and caulked. The annular overbreak surrounding the extrados of the lining was grouted.

The pioneering City & South London Railway, opened in 1890, was followed by the Waterloo & City Railway in 1898 and the Central London Railway, extending nearly 6 miles from Shepherd's Bush to the Bank, which opened in 1900. Deep-level tunnelling was now well proven, and the success of the Central London Railway led to the promotion of several new ventures. These were the Baker Street and Waterloo (*Bakerloo Line*), the Great Northern, Piccadilly and Brompton (*Piccadilly Line*) and the Charing Cross Euston & Hampstead (part of the *Northern Line*) Railways. Difficulty in raising capital delayed the construction of these projects, but the acquisition by an American financier, C. T. Yerkes, and his partners of the powers to build them saw all three completed across central London by 1907. In addition to these tube lines the Yerkes group took over the Metropolitan District Railway in 1901 and formed the Metropolitan District Electric Traction Co. Ltd. By 1902 this company had been absorbed into the Underground Electric Railways Company of London Ltd. and work had started on the construction of Lots Road power station to supply, first, the District Line (from February 1905) and, later, the Yerkes tube lines.

A mechanical excavator, working within a shield, designed by Thomas Thomson and a rotary excavator by John Price were tried out with limited success on the Central London Railway. Improvements to Price's design led to much more extensive and successful use of his machines on the Yerkes tubes.

No new deep-level tube railways were built across central London until the advent of the Victoria Line in the 1960s, but in the preceding half of the century the original lines were extended into the suburbs in stages. Much of this work was notable for the quality of the station architecture, especially the designs of Charles Holden for the Piccadilly and Northern Lines.

CROOME, D. F. and JACKSON, A. A. *Rails Through the Clay*. Capital Transport, Harrow Weald, 2nd edn, 1993.

BARKER, T. C. and ROBBINS, M. *A History of London Transport*. Vol. 1, *The Nineteenth Century*, 1963; vol. 2, *The Twentieth Century to 1970*. Allen & Unwin, London, 1974.

26. City & South London Railway

HEW 2277

Except for the short-lived and small-scale Tower Subway, opened in 1870, this important undertaking was the world's first *tube* railway, built using a tunnelling shield, and also the first urban electric railway in England. In 1884 an Act (47 & 48 Vict. c.167) was obtained to build the *City & Southwark Subway*. The original plan to have cable-hauled trains worked by stationary steam engines was discarded and the motive power was provided by small four-wheel electric locomotives. The use of electric traction and the availability of safe hydraulically operated passenger lifts had made deep-level underground railways a practical possibility. The engineers involved were P. W. Barlow and J. H. Greathead, who had collaborated on the Tower Subway. Sir John Fowler and Sir Benjamin Baker acted as consulting engineers. Construction began in October 1886 and an Act of 1887 empowered its extension to Stockwell. The northern terminus was *King William Street* in the City. The river tunnels were undertaken first from a shaft near the Old Swan Pier. Further shafts were sunk at the Monument, St. George's Church, the Elephant & Castle and Kennington. The railway was formally opened by HRH the Prince of Wales on 4 November 1890 and public service began on 18 December. In the same year powers were obtained to extend the subway to Clapham, and to change its name to the *City & South London Railway*.

The internal diameter of the tunnels on the City & South London Railway and its early extensions varied from 10 ft 2 in. to 11 ft 6 in. During 1922–24 these tunnels were rebuilt to the standard dimensions that had been established on the tube system (11 ft 8¼ in. on the straight and 12 ft 6 in. on curves). At the same time the Camden Town junctions were built to link the City & South London Railway to the Hampstead line.

CROOME, D. F. and JACKSON, A. A. *Rails Through the Clay*. Capital Transport, Harrow Weald, 2nd edn, 1993.

BARKER, T. C. and ROBBINS, M. *A History of London Transport*. Vol. 1, *The Nineteenth Century*, 1963; vol. 2, *The Twentieth Century to 1970*. Allen & Unwin, London, 1974.

27. Post Office Railway

HEW 222

This private underground train system has carried mail across London since 1927. The railway is 6½ miles long, running between *Paddington* and *Whitechapel*, and serves two main line rail stations (*Paddington* and

Liverpool Street) and six sorting offices. The railway resulted from an Inquiry of 1911 into the problems of moving mail through crowded streets. The Engineer was H. H. Dalrymple-Hay. Work started in 1913, but was interrupted by the First World War. The tunnels were completed in 1926 and were then handed over to the English Electric Company for fitting out.

The trains run on 2 ft gauge track in 9 ft diameter cast-iron lined tunnels between stations. The average depth of the railway is 70 ft, but stations are nearer the surface. These two-way tunnels divide at station approaches into two 7 ft tunnels each carrying one track. The cars are 27 ft long, each has four containers which can carry 15 bags of letters or six bags of parcels. The railway runs 22 hours a day, maintenance work being carried out between 8 a.m. and 10 a.m. The railway uses a central third-rail traction current system. In station areas the third rail carries 150 V DC, giving a speed of 7 miles per hour, or 440 V DC in main tunnels, giving a speed of 35 miles per hour. At peak times there is a train each way every four minutes, while at other times there are 12 trains an hour in each direction.

28. Victoria Line

HEW 2278
TQ 373 890 to
TQ 312 756

During the late 1940s London Transport was able to put in hand the completion of several schemes that had been halted by the Second World War, such as the Central Line extensions and, on the Metropolitan Line, quadrupling the tracks from Harrow to Moor Park and extending electrification to Amersham and Chesham.

Provision for building the Victoria Line from Victoria to Walthamstow was included in the British Transport Commission Act of 1955. Major objectives were to provide extra passenger capacity on a north-east/south-west route traversing the West End, and to link the main line railway stations at Victoria, Euston, King's Cross and Finsbury Park. An experimental length of about a mile of twin tunnels on the proposed line of the railway was built in the Finsbury Park area in 1960–61, but it was not until August 1962 that Government approval was given for the full project.

Largely because of a shortage of iron and steel during the rearmament programme of the late 1930s, bolted precast reinforced-concrete segments had been used to line some of the tunnels on the extensions to the Central Line. A further innovation on the Victoria Line was the use of unbolted concrete and cast-iron lining segments in the running tunnels, where these were driven through stiff, dry, London Clay. Each ring of lining was assembled in position and was expanded against the surrounding clay by jacks inserted into pockets in the ring of segments. The jacks were then replaced by packing pieces. The principle of this so-called 'flexible' lining design and installation procedure is that the passive ground resistance is mobilised to exert virtually uniform radial pressure around the ring of segments. Slight deformation may occur but there is no requirement to develop resistance to bending. With no annular space to be grouted and no bolting required it is possible to build the rings much more quickly than when bolted segments are used.

Four each of two different designs of tunnel boring machines were used to drive the running tunnels on the Victoria Line. For the much larger station tunnels Greathead shields were employed, the excavation being carried out by hand mining, and the linings generally being formed from bolted cast-iron segments, except in a few cases where steel linings were installed to carry superimposed point loads. To deal with unstable ground conditions, particularly where the line traverses the Woolwich and Reading Beds, compressed air working, ground freezing and chemical consolidation were variously used during construction.

Most of the tunnelling between Walthamstow and Victoria was carried out during 1963–66. After fitting out, the line was opened in stages, first from Walthamstow to Highbury & Islington on 1 September 1968, next to *Warren Street* on 1 December, and finally to Victoria on 7 March 1969. This part of the line, about 10¼ miles long, cost just under £71 million, including £7.45 million for rolling stock. The 3½ mile extension to Brixton did not receive Government approval until August 1967. Built in generally more difficult ground than most of the Walthamstow to Victoria section had been, the additional length was hand mined using Greathead shields. Opened on 23 July 1971 it cost about £21½ million. Trains on the Victoria Line are operated automatically, with provision for manual operation if circumstances require it.

CROOME, D. F. and JACKSON, A. A. *Rails Through the Clay*. Capital Transport, Harrow Weald, 2nd edn, 1993.

HORNE, M. A. C. *The Victoria Line*. Douglas Rose, London, 1988.

FOLLENFANT, H. G. *et al*. The Victoria Line. *Proc ICE*, Suppl., 1969, Paper 72705; Discussion, *Proc. ICE*, Suppl., 1970.

DUNTON, C. E., KELL, J. and MORGAN H. D. Victoria Line experimentation, design programming and early progress. *Min. Proc. Instn Civ. Engrs*, 1965, **31**, 1–24.

29. Docklands Light Railway

HEW 2272

The regeneration of the London Docklands necessitated a major investment to improve public transport services throughout the area. In 1982 the Government agreed to fund, through the London Docklands Development Corporation and jointly with the Greater London Council, a light railway extending westward from the City (TQ 327 813) to North Quay, near Poplar, whence branches would run north to *Stratford* (TQ 386 844) and south across the Isle of Dogs, terminating at *Island Gardens* just north of the Thames. A light railway was seen at first as a less costly alternative to a suggested extension of London Underground's Jubilee Line, but eventually both were built.

The Docklands Light Railway Act received Royal Assent in April 1984. Design and construct contracts were let in August 1984 for the line from Tower Gateway to Island Gardens and, in April 1985, from North Quay to Stratford. These lines were opened by the Queen on 30 July 1987 and the first fare-paying passengers were carried a month later. Meanwhile, in November 1986 Royal Assent was granted for an extension to Bank Station on the London Underground system. Tunnelling for this section began in March 1988. In July 1989 a proposed extension eastwards from Poplar to *Beckton* (TQ 432 815) received the Royal Assent and construction started in January 1990. The lines to Bank and Beckton were opened in 1991 and 1994, respectively. Royal Assent was granted in 1993 for an extension of the line southwards from Island Gardens to *Lewisham* (TQ 375 736) . Work started in October 1996 and the line was opened in November 1999.

A proposed branch from Canning Town to North Woolwich serving the London City Airport is under consideration.

Much of the 12.1 km route first built, and opened in 1987, made use of pre-existing railway formation and viaducts. It follows the line of the London & Blackwall Railway (see HEW 2271, p. 161) from the City to Poplar and its branch to Bow and thence to Stratford via the former Eastern Counties line. Southwards from Poplar the railway, including three of its stations, is carried over the West India Dock on a new 16-span viaduct, 610 m long, of composite construction (reinforced-concrete deck slabs on steel girders resting on reinforced-concrete piers). It then picks up the formation of the former Millwall Extension Railway. This alignment was followed above ground to a terminal at Island Gardens, until this station was replaced by a subsurface one on a new alignment when the tunnels under the Thames were built for the Lewisham extension. The tunnels comprise cut-and-cover approaches to twin bores, each 1.1 km long, driven through water-bearing strata. A tunnel boring machine incorporating a pressurised slurry bulkhead chamber to support the excavated tunnel face was used. The bored tunnel lining is formed of precast concrete segments and has an internal diameter of 5.2 m. Other interesting civil engineering works on the Lewisham extension are the *Cutty Sark* subsurface station, designed to resist flotation by its own dead weight, and the 20-span sinuous viaduct which winds its way for 800 m across Deptford Creek.

The passenger vehicles are single cars 28 m long, articulated at mid-length, enabling them to negotiate curves of 40 m radius, such as those through the turnouts at the delta junction at North Quay. Traction current is supplied at 750 V DC, collected from the underside of a third rail housed in a shroud protecting its top and side surfaces.

Train control was fully automatic from the time of opening. The line to Beckton was equipped with a moving block system—the first in the UK—which was subsequently extended to cover the whole railway.

PEARCE, A., HARDY, B. and STANARD, C. *Docklands Light Rail Official Handbook*. Capital Transport, Harrow Weald, 2000, 4th edn.

30. Jubilee Line Extension

HEW 2279
TQ 289 804 to TQ 386 844

London Transport's Jubilee Line began as an extension from Baker Street to Charing Cross, of the Stanmore branch of the Bakerloo Line. It was initially called the Fleet Line. Parliamentary powers were obtained in July 1969 and construction began in 1971. A target date of 1977 was set for opening and the line was renamed the Jubilee Line to commemorate the Silver Jubilee of HM Queen Elizabeth II. In the event, it did not open until 1 May 1979.

Meanwhile possible routes for a further extension into the London Docklands area were being considered, but the Bill for the 16 km route from Green Park to Stratford was not deposited until 1989, with Royal Assent being given in March 1992. Construction began in the last quarter of 1993.

Except for 3.5 km at its eastern end the extension is in tunnel. From Green Park to London Bridge it traverses London Clay, but most of the length between London Bridge and Greenwich lies in the water-bearing Thanet Sands or Woolwich and Reading Beds. The line crosses under the Thames at four places.

Of the 11 new stations all but two provide interchanges with other lines. Passenger flows arising from these facilities and from the expected further development of the Docklands led to several of the stations being built on a much larger scale than has been seen

Jubiliee Line extension, completed running tunnel between Canada Water and Canary Wharf

QA PHOTOS LTD

Westminster Station, Jubilee Line extension

E. C. DIXON

previously on the tube railway system. Those at Westminster, Canada Water, Canary Wharf and Greenwich are particularly impressive.

Difficult structural problems had to be overcome at many sites where existing underground lines and complex arrays of foundations of other buildings were already present. At Westminster provision had to be made for a new parliamentary building to be built directly over the station. A bold and imaginative gesture on the part of London Underground led to the appointment of a number of different architects and structural engineering consultants who were given a free hand to produce individual designs for each station. The result was a series of structures of outstanding originality and interest.

A range of techniques was used to cope with the variety of strata encountered when driving the running tunnels. The New Austrian Tunnelling Method (NATM), involving the use of sprayed concrete ('shotcrete') to stabilise newly excavated surfaces, was employed in some sections running through London Clay. Compressed air working was necessary in the water-bearing strata, and tunnel boring machines specially designed to support the unstable exposed surfaces as excavation proceeded were used.

Automatic train operation with moving block signalling was originally envisaged for the Jubilee Line, but it proved impossible to develop a satisfactory system within the planned construction schedule and the scheme was abandoned in favour of established train working technology.

FIELD, C., GAMBLE, M. and KARAKASHIAN, M. Design and construction of London Bridge Station on the Jubilee Line Extension. *Civil Engrng*, 2000, **138**, 26–39.

New Civil Engr, Supplements, February 1994, September 1996, February 1997, October 1999.

31. Lyne Railway Bridge

HEW 2280
TQ 022 670

Opened to rail traffic in February 1979, this interesting bridge carries the double-track line between Chertsey and Virginia Water over the M25 motorway. It is the only cable-stayed underline railway bridge in the United Kingdom.

The pronounced skew and limited construction depth available led to a design comprising a channel formed from two cable-stayed prestressed concrete edge beams 2760 mm deep and 1200 mm wide, continuous over centre supports, with a 600 mm thick reinforced-concrete deck slab. Each span is 54.87 m between bearing centres. Two reinforced-concrete towers are cantilevered from the edge beams above the central supports and are 24.8 m high from the deck soffit. Four pairs of stays from each tower are anchored at the third points of the edge beams.

The bridge was built in its permanent position in advance of the motorway works, the tracks being slewed clear of the working site to enable trains to continue running during construction. The unusual design was the result of a collaboration between Stressed Concrete Design Ltd. and Mr. J. Bucknall, Bridge Engineer, British Railways, Southern Region. Redpath Dorman Long Contracting Ltd. was the main contractor.

KRETSIS, K. Lyne Railway underbridge over the M25. *Proc. ICE*, 1982, **72**, Pt. 1, 585–610.

Concrete Quarterly, 1981, **128**, 36–39.

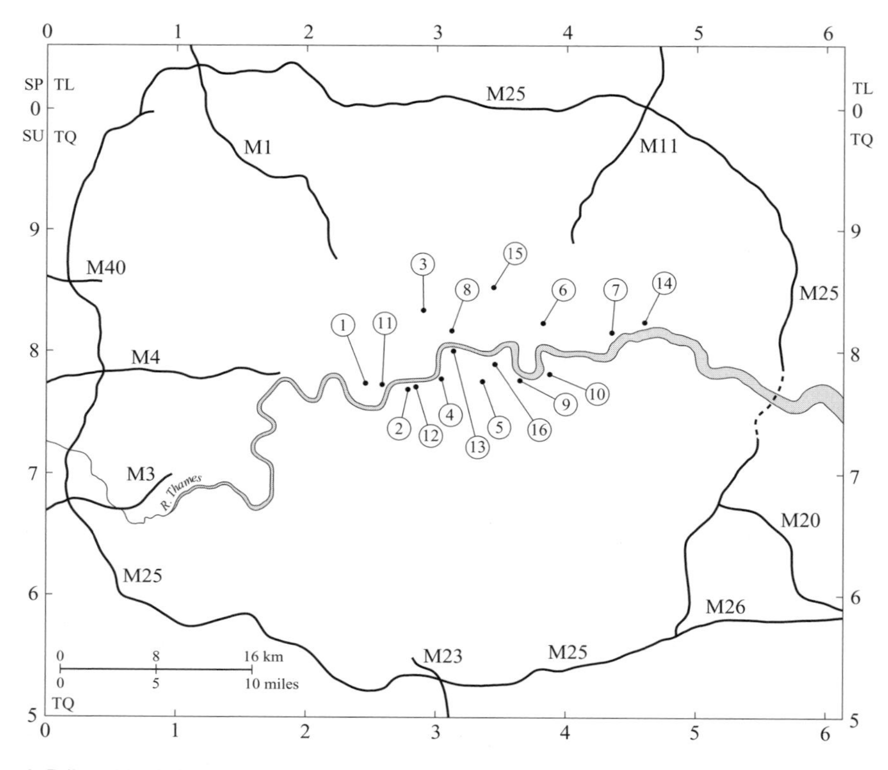

1. Fulham No. 2 Gasholder
2. Gasholders at Battersea
3. St. Pancras Gasholders
4. Kennington Gasholders
5. Old Kent Road Gasholders
6. Bromley-by-Bow Gasholders
7. Beckton Gasworks
8. Holborn Viaduct Power Station
9. Deptford Power Station
10. Greenwich Power Station
11. Lots Road Power Station
12. Battersea Power Station
13. Bankside Power Station
14. Barking Power Station
15. Wapping Pumping Station
16. Rotherhithe Pumping Station

7. Energy Generation and Distribution

Urban living makes essential demands for energy to sustain its operations by day and night, throughout the year. These demands include energy for such purposes as lighting and heating, and power to maintain public utilities, manufacturing, transport and communications systems. Although London had for centuries harnessed the energy of animals, water and, to a lesser extent, wind, for milling and pumping purposes, it also developed a great demand for coal. This had to be transported to London, largely from the north-eastern coalfields by sea and, increasingly from the middle of the nineteenth century, by rail. The coal supply industry has left its mark on the industrial urban landscape. Today the Thames riverside still has a heritage of cranes and disused jetties, in timber, cast iron and reinforced concrete, relating to the coal-supply industry serving gasworks and power stations, and fleets of colliers were a common sight on the river. Seaborne coal supplied many London industries, notably the water supply, gas, electricity, main drainage, hydraulic power and many other steam-powered industries. In addition, the duties levied on coal were a major source of finance for developing the infrastructure of the capital. The water-supply and gas- and electricity-generation industries shared common patterns of development. These industries started by establishing multiple, small-scale operating units often serving just one client, and from these small beginnings developed into large network systems largely achieved by amalgamation of private companies. The water-supply industry has always been among the first to adopt innovative technologies for pumping purposes. Waterwheels had been used under the arches of Old London Bridge from the sixteenth century, and at the end of the seventeenth century Captain Thomas Savery's *engine for raising water by the impellent force of fire* was built in London and tried out at the Thames-side York Buildings Waterworks. Thomas Newcomen's beam engine was also eagerly adopted in the early eighteenth century, and eventually James Watt's steam engine was almost universally used throughout the nineteenth century. Many of these survived in steam in London as the principal working engines in water and main drainage pumping stations until the 1950s and a little beyond. As an example of the enormous quantity of coal brought to London for the public utilities the Minute Books of the Metropolitan Water Board for 1930 record that in that year they alone used 198 729 tons of coal.

Gas

Eighteenth-century experimenters were aware of the flammable gas given off when coal is heated in a retort, and the first gaslighting was installed by William Murdock in 1792 at his home in Cornwall and in 1802 at Boulton & Watt's foundry at Soho in Birmingham. The commercial production of Town's Gas in London began early in the nineteenth century when the Chartered Gas Light and Coke Company laid mains under the streets of Westminster in 1813. It was to become a very large industry and first exploited gas extraction from coal as a means of lighting streets and premises. Gasworks were sited to facilitate the delivery of vast quantities of waterborne coal, and this led to the construction of river jetties, docks and coal-handling machinery. The industry's development was characterised by an early phase of a multiplicity of small works, each serving a limited area, followed by stages of amalgamation and eventually by nationalisation. A large network of gas mains and associated storage holders was built, enabling an enormous growth in industrial and domestic use of gas for lighting and, later, heating and power. Many London industries such as glassworks and potteries, which had previously used coke-fired kilns, changed over to gas. Large gasholders are distinctive features in the industrial landscape of London, and London has the greatest heritage of these in the country. The principles of low-pressure gas storage were soon established. The gas was stored in a cylindrical wrought-iron vessel called the *bell* with an open bottom immersed in a tank of water to seal in the gas. The bell moved up and down within vertical guides, to vary the volume of gas being stored. The early gasholders were *column guided*; later the ring of guide columns was replaced by helical guidance and the gasholders were described as *spiral guided*. The last generation of gasholders comprised fixed cylindrical structures with an internal piston to vary the storage capacity, and these were sometimes described as *piston* or *waterless* holders. Through most of the nineteenth century the London gas companies led the industry both in scale of operation and in technology.

Gas engineering in its early years was an important branch of civil engineering; pioneers such as Samuel Clegg were early members of the Institution of Civil Engineers (ICE) and Thomas Hawksley, who was President of the ICE, was first President of the Institution of Gas Engineers.

MALCOLM TUCKER

Gasholder, Kensal Green

Electricity

Electricity generation emerged in the late nineteenth century, and from the 1870s electric lighting became a serious competitor to the gas industry. In 1876 the carbon arc lamp arrived, and between 1877 and 1879 this was used to light the Gaiety Theatre (the first in London to adopt it),

Billingsgate Market, the Victoria Embankment, Holborn Viaduct and the Mansion House. The early power stations used steam or gas-engine driven dynamos generating *direct current* electricity through rope or belt drives. The carbon arc light became widely used in theatres and other large public buildings, such as railway stations and industrial premises, and civil engineering contractors also adopted it to extend their working hours on site. The world's first experiments in adapting it for lighthouse illumination were made by Michael Faraday and James Walker in the surviving small experimental lighthouse at Trinity Buoy Wharf at Bow Creek in the 1850s.

The incandescent electric lamp was first used in London in December 1881 at the Savoy Theatre. In 1882, an Electrical Exhibition at the Crystal Palace in Sydenham provided the wider general public with the first opportunity of seeing incandescent lamps in operation. A year later Sir Coutts Lindsay built London's first permanent generating station to light his Grosvenor Gallery in Bond Street. Three steam engines drove two Siemens alternators. These proved difficult to work and Lindsay invited the 21-year-old engineer S. Z. Ferranti to take charge of the generating station, a move which was to have momentous consequences. Ferranti made many changes in the plant and within three years the Grosvenor Gallery power station was supplying AC electricity to 100 miles of streets from Regent's Park to the Thames and from *Knightsbridge* to the Law Courts.

The generating industry quickly developed its own internal system of competition between two rival systems. Firstly, *direct current* supply, largely spearheaded by Colonel R. B. Crompton; and, secondly, *alternating current* electricity, generated by alternators and vigorously promoted by S. Z. Ferranti. This battle of systems, each hoping to become the industry standard, is similar to the railway gauge issue and is typical of new technologies to the present day. An early survival of a Crompton DC generating station site is at Kensington Court; this was opened in 1887 to serve a small development of apartments. Just two years later Ferranti's Deptford power station was opened—this was the flagship of AC generation, and in many ways the real beginnings of the large-scale generating and distribution industry.

In addition to public supply generating stations, several railway companies built their own electric power stations in London. For example, the Great Western Railway built its own generating station, which was opened in April 1886. It was sited between Paddington and Westbourne Grove stations, on the south side of the tracks, and was designed to light the Great Western Hotel, the offices and platforms at Paddington, neighbouring goods yards, together with the stations at Royal Oak and Westbourne Park. All had been previously lit by gas made by the Great

Western Railway. The original power station was closed in 1907 and replaced by one at Park Royal. The City & South London Railway had a power station at Stockwell, which started operating in 1890.

Hydraulic Power

The term 'hydraulic power' came into use in the middle of the nineteenth century to describe engineering power systems where the water was significantly above atmospheric pressure, therefore excluding traditional waterwheel technology. Three British engineers, Joseph Bramah, William Armstrong and the civil engineer Edward Bayzand Ellington, made pioneering contributions to the subject. Ellington's particular contribution was for the use of hydraulic power in urban distribution networks. Having pioneered Britain's first hydraulic power system in 1877 at Hull, he then turned his attention to London. In 1882 the General Hydraulic Power Company was formed, and in 1884 an Act (47 & 48 Vict. c.72) created The London Hydraulic Power Company. The system comprised a series of hydraulic pumping stations delivering water at 700 lbf/sq. in. into a network of cast-iron pressure mains north and south of the river. The advantage of the network was that clients did not have to generate their own energy but only had to be connected to the mains to have a clean source of energy delivered to their premises. The company supplied energy for cranes, capstans, lifts in hotels, Pelton wheels driving workshop machinery, revolving stages and safety curtains in theatres, and organ blowers in churches. Most of the machinery was provided by the Hydraulic Engineering Company of Chester, of which Ellington was a Director. Ellington took on Corbet Woodall as a partner, and as consulting engineers they designed schemes for London, Liverpool (1884), Melbourne (1889), Birmingham (1891), Sydney (1891), Manchester (1894) and Glasgow (1895). The London Hydraulic Power Company provided a design, manufacture, installation and maintenance service, and the clients paid for their metered water on a sliding-scale tariff. At the same time major industries, such as gasworks, railway and dock companies, and other industries would often have their own internal hydraulic systems, engineered by Armstrong or other engineering firms.

The first London Hydraulic Power station was opened at Falcon Wharf on Bankside, in 1883, serving just 7 miles of mains. The system eventually comprised eight pumping, or hydraulic accumulator, sites, including Kensington Court (1883), Phillip Lane (1884), Millbank (1888), Wapping (1892), City Road (1893), Rotherhithe (1903) and Grosvenor Road (1910). The company developed and reached its peak in the late 1920s and early 1930s. In 1928 they sold 1.7 million gallons of pressurised water serving 8005 hydraulic machines connected to the mains. In 1930 the company burnt nearly 22 000 tons of coal, a figure never to be repeated, and the mains reached a maximum length of 186 miles in 1935. The mains crossed the Thames over Southwark, Waterloo and Vauxhall bridges, and later also used the Tower Subway (HEW 233). Despite the obvious advantages of hydraulic power, competition from electricity led to a gradual running down of the system. Electrification of the machinery in all the pumping stations from 1953 resulted in the total loss of the company's steam plant. The last pumping station was closed in 1977.

TUCKER, M. T. *London Gasholders Survey*. English Heritage Report, December 2000.

PARSONS, R. H. *The Early Days of the Power Station Industry*. Babcock and Wilcox, Cambridge, 1939.

PREECE, A. H. The electricity supply of London. *Min. Proc. Instn Civ. Engrs*, 1898, **134**, 121–205.

ELLINGTON, E. B. Hydraulic power supply in London. *Min. Proc. Instn Civ. Engrs*, 1893, **115**, 220–41.

ELLINGTON, E. B. The distribution of hydraulic power in London. *Min. Proc. Instn Civ. Engrs*, 1887–88, **94**, 1–30.

Gas Supply

1. Fulham No. 2 Gasholder

HEW 95
TQ 260 769

John Kirkham, the engineer of the Imperial Gas Light and Coke Company, one of the first large gas companies, built this gasholder in 1830. It was in use until 1971 and is now by far the oldest gasholder in existence. The nominal capacity of this holder was 230 000 cu. ft. Although very small compared with the modern gasholders around it, it was of record size in its day, being 100 ft in diameter and 30 ft high, in a single lift, within a brick tank of similar dimensions set in the ground. The bell is of light, riveted, sheet metal, and to support its gently domed crown when the gas pressure was released 36 radial trusses of wrought-iron bars span to a central cast-iron king post landed on the dumpling in the tank. Stability is provided by 12 cast-iron standards around the perimeter, of free-standing 'tripod' form with elegant roundels in the web. Flanged rollers on the top of the bell engage guide rails on the tripods in the usual way, but at the bottom there are antiquated cast-iron slippers encircling guide bars fixed to the tank.

No. 2 holder is not visible or accessible to the public. Fulham No. 7, of 1876–79 on *Imperial Road* (TQ 260 767), illustrates a revival of buttress-type guide standards.

WILSON, A. *London's Industrial Heritage*. David & Charles, Newton Abbott, 1967.

The Engineer, 1949, **188**, 459.

2. Gasholders at Battersea

HEW 2360
Centred at
TQ 288 771

Sandwiched between two railway viaducts on the approaches to Victoria Station is another, compact group of gasholders. Battersea No. 4 has a modern, spiral-guided bell (a system invented in 1887), but Battersea No. 5 of 1876–77 has a guide frame of giant, 60 ft tall cast-iron columns, styled in the Doric order, with a single girder round the top. This is the last remaining of a type commonly used up to that date for large holders. It has a very shallow untrussed crown, rising to 1/29th of its diameter, another rare period feature. This two-lift holder, holding 1½ million cu. ft, was designed by Robert Morton, MICE,

Gasholders, Battersea

MALCOLM TUCKER

of the London Gas Light Company. He also designed Battersea No. 6, holding 2.6 million cu. ft, in 1880–82. This is an accomplished early use of box-lattice wrought-iron standards (cf. Beckton No. 9), with two tiers of lattice girders with vertical webs and a box-lattice top girder.

Battersea No. 7 was erected for the Gas Light and Coke Co. in 1930–32 to the German, MAN design of waterless gasholders. Up to 6.6 million cu. ft of gas is contained by a piston, free to move up and down a cylinder 295 ft high, which rivals the chimneys of Battersea Power Station. There is another of this type at Southall (TQ 119 797).

3. St. Pancras Gasholders

HEW 2281
TQ 300 833

The Imperial Gas Light and Coke Company was formed in 1821 and produced gas on this site from 1822 to 1904. The site lies to the south of the Regent's Canal and just to the east of the Midland Mainline railway out of St. Pancras. By 1869 it was the largest gasworks in London. Three gasholders on the site, known as Nos. 10, 11 and 12, are of the so-called 'linked-triple' configuration as they share structural members, being tied together by short girders, a system unique in Britain. The holders are column guided with latticed circumferential girders at three levels. The southernmost holder (No. 10) was commissioned in May 1861, No. 11 in 1864 and No. 12 in 1867. The three holders are Listed Grade II. The Channel Tunnel Rail Bill of 1997 gave powers to London & Continental Railways to lay new track to the north of St. Pancras train shed, with threatened demolition of several nineteenth-century and twentieth-century gasholders. English Heritage has secured an alternative to demolition. London & Continental Railways have agreed to pay for their dismantling, storage and ultimate re-erection and use on a nearby site.

MIEHLE, C. The King's Cross gasholders. *Ind. Archaeol. News*, 1997, **103**, 2–3.

Gasholders, St. Pancras

WENDIE TEPPETT

Gasholder No. 1, Kennington

MALCOLM TUCKER

4. Kennington Gasholders

HEW 2373

Kennington No. 1 Gasholder (TQ 310 779) is a scenic feature of the Surrey County Cricket Ground at the Oval. When first built in 1877–79 by (Sir) Corbet Woodall, its 3.1 million cu. ft nominal capacity was the largest to date, but in 1890–91 it was heightened to four lifts and 6.1 million cu. ft by Frank Livesey, MICE.

The standards are of wrought-iron lattice work, T-sectioned with a strong taper, following the first of that type built at Hove in 1876. When increased in height to

133 ft, a vertical section was inserted, and the diagonal bracing was made stronger. The topmost telescopic lift is 'flying', i.e. it is able to rise some 40 ft above the limit of the guide rails on a new principle developed by Frank and his elder brother, George Livesey (see site 5 below). The tank is of brick, 218 ft diameter and 44 ft 6 in. deep, built tight against the London Clay in its lower part, upon undercut footings of concrete.

Close by there are Corbet Woodall's slightly earlier No. 4 and No. 5 Gasholders (TQ 309 780), each of 1 million cu. ft, built in 1873–74 and 1875–76. Their guide frames are joined like Siamese twins. The two tiers of cast-iron columns and girders resemble Bromley-by-Bow (see site 6, p. 207), but they are more dainty, structural progress being marked by light diagonal braces between columns and horizontal trussing bars (wind ties) around the top.

No. 2 Gasholder, now spiral-guided, is on the site of one built in 1854, which had remarkable tubular wrought-iron columns of 160 ft diameter.

LIVESEY, G. The guide framing of gasholders. *J. Gas Lighting*, 1888, **52**, 846–47, 887–90.

King's Treatise, 1879, **2**, 76–78, 131–36, 169–75.

WOODALL, C. Gasholders. *Proc. Br. Assoc. Gas Managers*, 1874, 37–43.

5. Old Kent Road Gasholders

HEW 2282

Close to the A2 *Dover Road*, in Peckham, there stands a prominent gasholder, 215 ft in diameter and 160 ft high, with a guide frame of sleek, clean lines that could reasonably be mistaken for a twentieth-century structure. But No. 13 Gasholder (TQ 348 777) was in fact built in 1879–81. Its great height, economy of construction, and capacity of 5.5 million cu. ft, broke new ground, the result of radical rethinking by its designer, (Sir) George Livesey, the Engineer to the South Metropolitan Gas Company.

Gasholder guide frames had previously been thought of as a series of individual cantilever standards, linked by the girders acting as catenaries to share the wind loading and perhaps stiffened from excessive sway by the addition of bracing, but not performing as a single entity. Livesey saw that the frame could behave like a stocky

MALCOLM TUCKER

Gas holders 12 and 13, Old Kent Road

beam of tubular cross-section, in tension on the windward side, and compression on the other, in effect a shell structure. The shear forces, greatest on the neutral axis, were to be transmitted by strong diagonal bracing. The standards became mere stiffening ribs, to distribute the wind loads applied to them by the bell and resist the vertical overturning forces. Rowland Ordish did the calculations and, after the Tay Bridge collapse inquiry, (Sir) Benjamin Baker gave a reassuring second opinion of its safety.

Every aspect was at the forefront of design. The wrought-iron frame is riveted throughout and the standards are of equal-flanged section with a solid web to ensure thorough painting and no rust traps. Slightly tapering, the standards have a maximum overall depth of 1/90th of their height, five times, or more, slimmer than previous practice. In place of girders there are slim struts of cruciform cross-section, in five tiers, while the wind bracing is of shallow, inclined, flat bars, with two sets in each panel to emphasise their important role.

More 'firsts' include the use of mild steel in the (since removed) top curb of the bell to resist the large shell forces from the untrussed crown, and tangential guide rollers in addition to the radial ones, to distribute the wind loads more effectively around the guide frame. The tank, 218 ft in diameter, has the exceptional depth of 55 ft 6 in. It is of mass concrete reinforced with embedded iron bands, and was cast directly against the dewatered and timbered face of the Thanet Sand to save on excavation and compacted backfill.

In No. 12 Gasholder (TQ 347 778) of 1874–75 Livesey achieved the considerable height of 92 ft. The slender standards are in cast iron, of I-section with decorative quatrefoils pierced in the web. There are three tiers of lattice girders with light wind bracing, and the top is stiffened by trussing with wire-rope 'wind ties'. At one time a 'flying' lift was added, another Livesey innovation (see site 4, p. 204). The tank, 184 ft in diameter and 47 ft deep, was the first large tank built entirely of mass concrete.

The tank of Gasholder No. 11, now filled in, had pioneered composite brick and concrete construction in 1872. At the rear of the present site there is Gasholder No. 10 (TQ 347 779) of 1866–67, which is of relatively conventional design, with cast-iron guide columns and a puddle-backed 36 ft deep brick tank. But, here, Livesey (with the contractor Thomas Docwra) mastered the art of dewatering the treacherous Thanet Sand, by sinking wells into the underlying chalk. This avoided the severe subsidence through loss of ground, previously experienced when pumping from sumps in this fine-grained sand.

LIVESEY, G. The principles of gasholder construction. *Trans. Gas Inst.*, 1882, 46–58.

The Engineer, 1880, **50**, 175.

J. Gas Lighting, 1880, 36.

LIVESEY, G. Gasholder tanks, difficulties and mistakes in their construction. *Proc. Br. Assoc. Gas Managers*, 1880.

King's Treatise, 1879, **2**, 78–79, 85–88, 146–49.

LIVESEY, G. Further experience of concrete tanks. *Proc. Br. Assoc. Gas Managers*, 1876, 16–21.

6. Bromley-by-Bow Gasholders

HEW 2359
TQ 386 823

These works were built by the Imperial Gas Light and Coke Company in response to a rival works recently built at Beckton. The Bromley works first produced gas in

MALCOLM TUCKER

Gas holders, Bromley-by-Bow

1873. The site occupies 160 acres bounded by the District Line railway, the River Lea and Bow Creek. Coal was delivered by barge from the tidal Bow Creek through a Lock into the Company's own Dock (TQ 386 819), which is 1000 ft long and 100 ft wide.

The remarkable group of seven (originally nine) large gasholders (TQ 386 826) of 1872–82 was commenced by Thomas Kirkham of the Imperial Gas Light and Coke Company, to the design of their Joseph Clark, and completed by Vitruvius Wyatt of the Chartered Gas Light and Coke Company, London's two largest gas companies having amalgamated meanwhile. They are all 200 ft in diameter with guide frames 70 ft high, to hold 2 million cu. ft of gas in two telescopic lifts. They have cast-iron classical columns in two orders, exquisitely detailed and with bold entablatures, and two tiers of composite girders of wrought-iron flanges and filigree cast-iron webs. The last four holders to be built have 24 columns rather than 28.

The road entrance to the works is via *Twelvetrees Crescent*, which is carried on a bridge over the River Lea and Bow Creek. The bridge (TQ 383 824), known as Bromley Bridge, was designed by P. W. Barlow and is a two-span structure with wrought-iron plate girders having arched soffits. Between the plate girders the bridge also carries two 48 in. diameter gas mains, which delivered gas to central London.

NEWBIGGING, T. and FEWTRELL, W. (eds.). *King's Treatise on Coal Gas*, vol. 2. Walter King, London, 1879.

7. Beckton Gasworks

HEW 2283

The Chartered Gas Light and Coke Company obtained powers to build the works at Beckton under an Act of Parliament in 1868 and work began in November 1868 to the designs of their Chief Engineer, Frederick Evans, MICE. The name 'Beckton' was chosen to honour Simon Adams Beck, the 'Governor' of the Company. The first pile of the river wall was driven on 29 November 1868 and gas was delivered for the first time in December 1870 through a 48 in. diameter main to the City of London. The site eventually occupied about 300 acres and the civil engineering contractor for the works was John Aird. Its site is on the open marshland of East Ham, and its Thames-side location allowed for the first time direct delivery of coal by sea from Durham and Northumberland. Still remaining are the concrete-filled cast-iron piers of the T-shaped jetty (TQ 449 813), where collier ships owned by the company were discharged by hydraulic cranes, and those of the second jetty added in 1895. From the jetty, conveyors and a railway system led to the retort houses. The original coaling plant was augmented by an eastward extension of the pier in July 1926. The extension is about 400 ft by 69 ft, constructed in prestressed concrete and carried on 268 piles. No. 2 Pier (TQ 447 810) was used for the export of residuals, notably coke. The last shipload of coal was delivered to the piers on 16 April 1969, and only some derelict reinforced-concrete coal bunkers now remain of the plant, but two gasholders are of interest.

No. 8 Gasholder (TQ 442 815) designed by Vitruvius Wyatt, built in 1876–79, contained 2 million cu. ft of gas in two telescopic lifts. The tank is 195 ft in diameter and 37 ft deep. No. 9 Gasholder (TQ 441 814) was built in 1890–92 and was designed by George Trewby, MICE, Wyatt's successor. It holds nominally 8 million cu. ft, and is one of the largest gasholders in the United Kingdom. It has a mass concrete tank 250 ft in diameter and 45 ft 6 in. deep. The bell is of four lifts. The guide frame, an early use of mild steel for this purpose, is 180 ft tall. There are 28 gently tapering standards of box-lattice construction.

Trans. Incorporated Instn Gas Engrs, 1892, 167–72.

TREWBY, G. C. System of unloading and storing coals at the Beckton Station of the Gas Light and Coke Co. *Min. Proc. Instn Civ. Engrs*, 1881–82, **69**, 318–19, plate 5.

NEWBIGGING, T. and FEWTRELL, W. (eds.). *King's Treatise on Coal Gas*, vol. 2. Walter King, London, 1879.

The Engineer, 1870, **29**, 61 ff.

Electricity Supply

8. Holborn Viaduct Power Station

HEW 2286
TQ 316 815

This small generating station was established within the structure of Holborn Viaduct. The venture was financed by the Edison Electric Light Company of London and was opened on 12 January 1882, seven months before Thomas Edison's Pearl Street power station in New York. The Holborn power station can claim to be the world's first public steam-powered electricity-generating station serving the needs of the private consumer in addition to its public lighting function. The plant comprised a Babcock & Wilcox water-tube boiler (the firm's first to be used for electricity generation), and a steam engine driving two of Edison's large dynamos delivering a 110 V DC supply, with a capacity to supply 3000 lamps. The entrance to the power station, and the company's office and showroom, at No. 57 Holborn Viaduct, was on the north side of the viaduct.

9. Deptford Power Station

HEW 2287
TQ 376 778

This was London's first major power station and was designed by Ferranti to generate AC electricity at 10 000 V. This appeared at the time to be far too dangerous a voltage to be generated in central London, and the Greenwich site was therefore chosen. The station was built for The London Electric Supply Corporation Ltd., formed in 1887 with an authorised capital of £1 million. Ferranti designed the works on a scale beyond anything yet attempted in Britain, or in any other country. Work began on the riverside site in April 1888, and despite the large foundations required to support the steam engines and four alternators each weighing 500 tons, the building was completed by midsummer 1889 and the power station was commissioned later that year. The main building was 210 ft long by 195 ft wide and was supported on a 4 ft thick mass concrete raft. Coal was delivered to a quay on Deptford Creek and was transported by rail to the boiler house, with its two squat chimneys, some 350 ft from the river. The real problem with the station lay with the high-voltage mains cables required to distribute the energy. The first set purchased was laid in the early part of 1889 and was a failure. Ferranti decided that the only way to achieve success was to design and manufacture his own cables. These were made in 20 ft lengths and with the insulation and jointing problems solved some of the cables remained in service until 1933. The cables crossed the Thames over Charing Cross, Cannon Street and Blackfriars railway bridges, and the Metropolitan & District Railway gave permission for the cables to be laid in their tunnels. Until 1904 Deptford had largely supplied electricity for lighting purposes, but in 1909 it supplied the suburban electric service of the London, Brighton & South Coast Railway.

Deptford West power station was opened in 1929, adjacent to the historic Ferranti station. A further station on the site, *Deptford East*, was completed in 1957.

RIDDING, A. *S.Z. de Ferranti, Pioneer of Electric Power*. HMSO, London, 1964, Science Museum booklet.

FERRANTI, G. *The Life and Letters of Sebastian Ziani De Ferranti*. Williams & Norgate, London, 1934.

10. Greenwich Power Station

HEW 2288
TQ 388 781

Standing in *Old Woolwich Road* on the banks of the Thames with a 240 ft river frontage this station was opened on 26 May 1906 to supply the London County Council's 68-mile network of tramways. The twin-bayed steel-framed main building is of stock brick with Portland stone dressings and is 475 ft by 195 ft and 80 ft tall. The original plant comprised 24 Yarrow boilers working

Greenwich Power Station

DENIS SMITH

at 180 lbf/sq. in., supplying four reciprocating steam engines driving flywheel-type alternators. The generator sets, rated at 3500 kW each, delivered three-phase current at 6600 V. Coal was delivered to the site from the Thames and off-loaded onto the massive coaling pier with 16 large cast-iron Doric columns filled with concrete, and a crane transferred it to a coal conveyor and moved it to bunkers over the west yard. The original building had two octagonal brick chimneys 182 ft high. By 1910 the station also contained four 5000 kW turbine generators, and this extension required two further chimneys which were to be 250 ft high. But, as the station is virtually on the Greenwich meridian, objections from the Royal Observatory led to the height of the new chimneys being limited to 182 ft. The original reciprocating steam engines were taken out by 1922 and replaced by steam turbine machinery. The Generating Station has been modernised in several stages: in 1930–32, 1939–44 and 1956–59; lastly, in 1967 it was decided to replace the plant with eight Rolls Royce *Avon* aero-engine gas turbines. These were installed in pairs, starting in 1969 and the work was completed in 1972. Each gas turbo-alternator set comprises a gas generator, a power (work) turbine, an alternator, a transformer and automatic control systems, and is able to supply an output of 14 700 kW. This is one of a few early power stations to remain in use, and is also an early example of a steel-framed building in Britain and has architectural merit.

The station was designed by J. H. Rider, the London County Council Tramways Electrical Engineer, and H. Lovatt Ltd. was the building contractor. The steelwork contractor for the north section was J. Westwood & Company of Millwall and that for the south section was E. C. & J. Keay Ltd. of Birmingham. Maurice Fitzmaurice, Chief Engineer to the London County Council, was responsible for the design and construction of the coaling pier.

GUILLERY, P. Greenwich generating station. *London's Ind. Archaeol.*, 2000, 7, 3–12.

11. Lots Road Power Station

HEW 2289
TQ 265 771

This steel-framed brick-clad power station was built in 1902–04 alongside Chelsea Creek to serve the Metropolitan District Railway, and the Baker Street and Waterloo, the Charing Cross, Euston and Hampstead, the Great Northern, Piccadilly and Brompton tube railways. With the rise in the demand for power the station was updated in the late 1920s and again in the early 1930s. After a thorough study of future power demands it was decided in 1962 to carry out extensive modernisation works. These was completed in 1969, the power being produced by six 30 000 kW steam turbine sets. A decision has been taken by London Underground Ltd. to close this station and to take power from the National Grid. At the time of writing the massive building with its twin brick chimneys remains as an impressive riverside feature. The power station was one of the earliest steel-framed buildings in the London area.

12. Battersea Power Station

HEW 2290
TQ 289 775

This prominent group of Thames-side buildings is a well-known landmark and was developed from the 1930s. The design of the external elevations was by Sir Giles Gilbert Scott for the London Power Co. Ltd. The authority of the Electricity Commissioners was obtained in October 1927. Access to the A-station was obtained from *Battersea Park Road*, from sidings off the Great Western Railway and by the river on a frontage of about 680 ft. The contract for the foundations was let in March 1929, steelwork erection began in October 1930 and was completed in March 1932, and the brickwork and concrete of the superstructure was finished by about the end of May 1933. The buildings are of brown brick with four fluted reinforced-concrete

Battersea Power Station

DENIS SMITH

chimneys The A-station was opened in 1933. The civil engineering design and construction supervision was by C. Seager Berry, MICE, Chief Civil Engineer to the London Power Company; John Mowlem & Co. Ltd. were the main building contractors, and the structural steelwork contractors were Sir Wm Arrol & Co. Ltd.

The B-station was opened between 1941 and 1944, with the last phase opening during 1951–53. The art deco control room has been mothballed until the long-term future of the building has been determined. The plant was also notable for its early combined heat and power system, known as the Pimlico District Heating Scheme, opened in July 1951, where hot water was taken in pipes under the Thames to the north bank to a development of residential flats where it was stored in a heat accumulator. The completed scheme served about 3200 flats and a population of around 11 000. The buildings are Listed Grade II and schemes are being discussed for their adaptive re-use as part of London's leisure industry.

DONKIN, B., MARGOLIS, A. E. and CARROTHERS, C. G. The Pimlico District Heating undertaking. *Min. Proc. Instn Civ. Engrs*, 1954, **3**, Pt. 1, 259.

BERRY, C. S. and DEAN, A. C. The constructional works of the Battersea Power Station of the London Power Company Limited. *Min. Proc. Instn Civ. Engrs*, 1934–35, **240**, Pt. 2, 37–73.

13. Bankside Power Station

HEW 2291
TQ 320 805

The power station at Bankside has its origins in the formation of the City of London Electric Lighting Company formed in 1891. The company inherited an earlier small plant on the site and power under the City of London company was first generated on 12 June 1891. The station always suffered difficult load conditions owing to sudden fogs on the Thames. From 1934 the station was operated under the Central Electricity Board as part of the National Grid.

The modern A-station on this site was opened in 1938. The post-Second World War B-station was opened in 1952, and because of its prominent central location was given superior architectural treatment by Sir Giles Gilbert Scott. The tall square chimney in brick and the recessed panelling of the walls of the building were designed to make it an acceptable companion to St.

WENDIE TEPPETT

Bankside Power Station

Paul's Cathedral on the north bank. It was of technical interest in being the first large power station to be designed specifically for oil-fired boilers.

The station has undergone an extensive adaptive reuse conversion by the Tate Gallery as the exhibition site for their modern art collection and it was opened in 2000 as the *Tate Modern*. Large sculptures are exhibited in the old turbine hall of the power station and the original overhead travelling crane, marked 'Arrol', is used for handling large exhibits. The Millennium Bridge provides pedestrian access from the City of London.

14. Barking Power Station

HEW 2292
TQ 465 817

The site at Barking was chosen because of the ease of delivery of waterborne coal. The site was planned to accommodate, eventually, three power stations, each of two sections. The A-station was built for the County of London Electric Supply Company, the first section of which was opened in 1925, and was the largest built in Britain as a complete station at one time.

The construction involved 250 000 cu. yd of earthworks, half a mile of main road, 7 miles of standard-gauge railway and a 100 ft deep half-mile cable tunnel under the Thames. Barking B-station opened in 1933 and the C-station in 1957.

SABBAGH, K. *Power into Art*. Allen Lane, London, 2000.

WILLIAMSON, J. Jetty and pump-house at Barking Power-Station. *Min. Proc. Instn Civ. Engrs*, 1927–28, **226**, Pt. 2, 135–52.

The London Hydraulic Power Company

15. Wapping Pumping Station

HEW 2293
TQ 354 806

Built in red brick in 1892 by John Mowlem & Co. this pumping station is located in *Wapping Wall*. It took its water from a well and the Shadwell Basin of the London Dock. Coal was delivered over the wall by hydraulic crane from the Shadwell Basin. All hydraulic pumping stations have a group of characteristic features, including a coal store, a boiler house and chimney, cast-iron sectional water tanks on the roof, an engine room with overhead travelling cranes, a filter house, an accumulator tower and a house for the engineer. The original plant comprised six triple-expansion steam engines supplied by six Fairbairn–Beeley boilers. In addition, this

Wapping Pumping Station

WENDIE TEPPETT

pumping station has an underground reservoir, storing 420 000 gallons of water, with a brick jack-arched roof supported on cast-iron girders and columns. In 1923 the steam engines were replaced by two Mather & Platt electric centrifugal pump sets and the boilers were also removed at this time. New boilers and a Parsons steam turbine were in operation in 1926. During the 1950s and with the passing of the Clean Air Acts the steam plant was replaced by electrically driven three-throw ram pumps. This was the last station to remain in use, its life extended to supply Tower Bridge during its own conversion from steam to electricity, and the pumping was closed in July 1977. The Engineer for this, and other London Hydraulic Power Company stations, was Edward Bayzand Ellington. The station is Listed Grade II* and various options for the future use of the building have been discussed.

16. Rotherhithe Pumping Station

HEW 2294
TQ 354 795

Opened in 1903 this station is, unusually for the company, built in London stock brick. It is situated in *Renforth Street*. Coal was delivered by barge to the Albion Dock of the Surrey Commercial group, and loaded into wagons on a narrow-gauge railway, which transferred the fuel to the coal store. The station's water supply was also abstracted from the Albion Dock. This station has the usual array of features, including an octagonal brick chimney, an engine room with ridge ventilators, a boiler house with water tanks on the roof, an accumulator tower, a station engineer's house and an extensive maintenance workshop with machines powered by Pelton wheel turbines. The station was electrified in the 1950s, but was closed in 1977, when the engines and pumps were removed. The building is Listed Grade II and it is likely to be converted into dwellings.

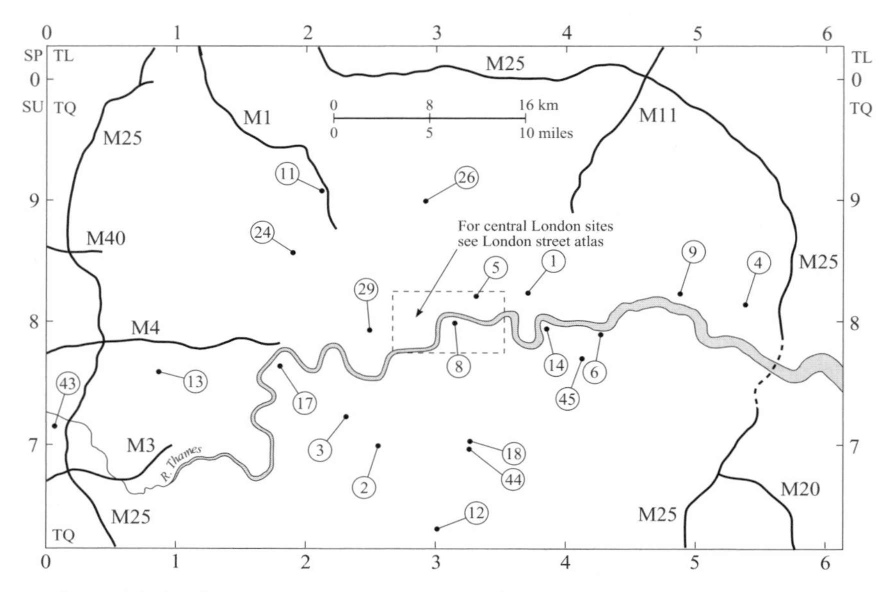

1. Three Mills Distillery
2. Liberty's Printworks, Merton
3. Wimbledon Common Windmill
4. Upminster Smock Mill
5. Whitechapel Bell Foundry
6. Royal Arsenal, Woolwich
7. Whitbread Brewery, Chiswell Street
8. Kirkaldy Testing Works, Southwark Street
9. Ford Motor Company, Dagenham Works
10. Trinity Buoy Wharf
11. Hendon Aerodrome
12. Croydon Airport
13. Heathrow Airport
14. Enderby's Wharf, Greenwich
15. Post Office (British Telecom) Tower
16. British Museum and Reading Room Roof
17. Palm House, Royal Botanic Gardens, Kew
18. Crystal Palace
19. The German Gymnasium
20. Bethnal Green Museum
21. Royal Albert Hall
22. South Kensington Subway
23. Olympia Exhibition Hall
24. British Empire Exhibition Site, Wembley
25. Royal Agricultural Hall, Islington
26. Alexandra Palace, Muswell Hill
27. Ritz Hotel, Piccadilly
28. Dorchester Hotel
29. Commonwealth Institute, Holland Park
30. Millennium Dome
31. British Airways London Eye
32. Millbank Tower
33. National Westminster Tower
34. Hop and Malt Exchange, Southwark Street
35. Smithfield Market
36. Michelin Building
37. Houses of Parliament
38. Westminster Hall Roof
39. Henry VIII's Wine Cellar, Whitehall
40. Trafalgar Square Waterworks
41. Cabinet War Rooms, Whitehall
42. Institution of Civil Engineers, Great George Street, Westminster
43. Royal Indian Engineering College, Cooper's Hill
44. Crystal Palace School of Practical Engineering, Sydenham
45. Royal Military Academy, Woolwich
46. University College, London

8. Notable Buildings

London has a magnificent heritage of historic buildings, but this chapter is largely based on structures that are of historic engineering significance. For description here we have made a selection of buildings which illustrate a development in structural form, the use of a new material, a technical advance in machinery or building services, or an association with eminent architects, engineers or contractors.

The categories of building use in London are extremely wide. They could be listed under the headings of structures built for Government purposes, ecclesiastical buildings, industrial buildings, those for conducting trade and commerce, the entertainment and tourist industry, buildings to facilitate transport and communications, and those associated with education and the professions. Civil engineers have been involved with all these categories. In addition, London's structural heritage covers a period going back to the Romans.

The acquisition of land and buildings designed, built and maintained for the Crown or Government, has certainly since the Tudor period required organisation, construction supervision and control of expenditure. The Office of the King's Works was for centuries controlled by a Board chaired by the Surveyor General, and comprising the Master Mason, the Master Carpenter and the Comptroller (or finance officer). The Surveyor General was usually an eminent architect, and the best known of the early period was Inigo Jones who held the office from 1615 to 1643. Other notable Surveyor Generals were James Wyatt and John Nash. The Office of Works therefore had its origins in the maintenance of the Royal Palaces, houses, parks and roads, and employed architects and engineers, but was, naturally, subject to political control. In the nineteenth century another body known as The Commissioners of Woods, Forests, and Land Revenues was formed, its functions somewhat overlapping with those of the Office of Works. Their office was, in 1851, at No. 1 Whitehall Place, and their jurisdiction was described as having 'the care of the national property, and the direction of public works and buildings not under the Admiralty or Ordnance'. The Woods and Forests had both revenue-producing and spending functions, which caused difficulties from time to time when revenues were used for public works without Parliamentary sanction. During the nineteenth century the Office of Woods and Forests expanded to include providing accommodation for Government departments. The Office of Works became responsible for the protection of ancient monuments in the late nineteenth century, and in 1940 the Office of Works became the Ministry of Works.

The status of London as the Capital City and the prestige attached to the seat of government of the world's largest empire naturally enabled clients to attract the attention of the most able architects and engineers who have worked in timber, masonry, iron and steel and concrete, often in a pioneering manner. The Great Fire in 1666 revealed the vulnerability of closely spaced timber buildings and led directly to the introduction of London's first building regulations. Fire regulations relating to buildings were strengthened by the Act of 1774 which was for 'the more effectually preventing Mischiefs by Fire within the Cities of London and Westminster ... and other Parishes', which was in turn superseded by the Act of 1865 'for the Establishment of a Fire Brigade within the Metropolis' which was to be run by the Metropolitan Board of Works.

Timber was an early structural material and was prized for its favourable strength/weight ratio and for its ability to sustain tension. It was widely used by contractors for such temporary structures as arch centring, scaffolding and cofferdams. Timber piles, notably of elm, were extensively used in foundation works of bridges and buildings. London has excellent examples of early timber roofs of hammer-beam construction, notably at Westminster Hall (see below), Eltham Palace and Crosby Hall. Timber was the ideal material for the interiors of dock warehouses, both for large-span roofs and also for flooring capable of taking heavy loading. The roofs of London theatres also required heavy carpentry—a good example is the Theatre Royal, *Drury Lane* where the roof trusses and under-stage machinery are largely of timber.

Cast iron, wrought iron and steel have been widely used in London for both structural and aesthetic reasons. The use of metal was often prompted by the need to improve the fire resistance of public buildings. Cast iron began to replace timber and masonry for columns in London buildings from the early nineteenth century, notably in dock warehouses and market buildings. Cast iron has often been used to mimic masonry, with decorative capitals to columns and elaborate screens, with excellent surviving examples in London pumping stations. Wrought iron, with its capacity to sustain tension, came to be used in large-span roof structures for railway stations and exhibition buildings. Rolled sections, first in wrought iron and later in steel, became widespread in the fabricated steelwork of framed structures.

Concrete was first used in buildings in the form mass concrete for footings, rafts, walls, arches and domes—structural forms in which tension is minimised. Notable examples in London included Robert Smirke's remedial foundation raft for the Millbank Penitentiary (1817), Joseph Tall's patent system of the 1860s for house and wall construction, examples of which survive in *Victoria Road*, Mortlake, and in *Woolwich Road*, Bexleyheath. A mass concrete church (1883) in *Waldegrave Road*, Crystal Palace, survives, and Westminster Cathedral (1903) has two 60 ft diameter mass concrete domes over the nave and one over the sanctuary. Many reinforced-concrete framed buildings in London were clad in masonry for aesthetic reasons or for prestige. Examples include the General Post Office building and the Central Hall, Westminster. Eventually, concrete structures developed an architectural aesthetic of their own and London has many good examples.

Although churches and cathedrals are in the care of the church and the surveyors of the fabric make annual checks of the structure, it is sometimes necessary to engage civil engineers to report on the condition. A notable example is that of St. Paul's Cathedral. In November 1912 the Dean consulted Sir Francis Fox who reported in 1913 revealing alarming conditions. The foundations were built on the rubble of the Great Fire and quicksands were also found. The dome was 5¾ in. out of plumb and the eight piers supporting it had subsided variously by between 2 in. and 6½ in.

London is not commonly perceived to be a manufacturing city, and yet, by the criteria of capital invested and the number of people employed, it was one of the largest manufacturing centres in the country. It is not surprising, therefore, that London has a rich heritage of industrial sites relating to water and windmills, and industrial premises of historical importance.

Similarly, the importance of trade and commerce to the Capital has left its mark on the built environment, including prestigious office blocks, exchange buildings dealing with corn, hops, coal and metal and, in the City of London, stocks and shares. There is also a heritage of specialist market buildings dealing with flowers, fruit and vegetables, fish and meat. The historical engineering works are arranged under headings grouped by the original use of the building.

ENGLISH HERITAGE. *Early Structural Steel in London Buildings*. Survey Report, October 2000.

THORNE, R. (ed.). *The Iron Revolution: Architects and Engineers and Structural Innovation 1780–1880*. RIBA Exhibition Essays, RIBA, London, June–July 1990.

Mills

1. Three Mills Distillery

HEW 2295
TQ 383 828

The two Georgian tidemill buildings in *Three Mill Lane*, the *House Mill* and the *Clock Mill*, comprise one of the most important industrial heritage sites, not only in London but also in Britain. There have been mills on this site since the Domesday Survey of 1086. Over the centuries the mills had been used for corn milling, and from the 1730s they were given over to the milling of grain for the distillation of gin. The buildings can be reached from the Three Mills Lane bridge over the River Lea. Speaking in 1938, Mr. W. M. M. Sheppard, the Chief Engineer of Three Mills, said that the waterwheels ran at 16 rev./min, with a working head of 10–12 ft and they reckoned to get 10–12 hp on each stone; the tide ran the mill for 7–8 hours, the gear teeth lasted for 40–50 years, and they employed one millwright and four carpenters.

The House Mill is the oldest building surviving on this site. It was built in 1776 by Daniel Bisson, whose plaque survives on the south elevation. The building is of five storeys, two of which are within the twin-ridged roof. The building has a brick southern wall and a timber framed

Three Mills Distillery

DENIS SMITH

weatherboarded north elevation. The timber ground floor is carried on cast-iron beams spanning the mill race. The first and second floors comprise north–south aligned timber beams supported by cast-iron columns. The third and fourth floors are entirely of timber posts, beams and floor boarding. Remains of the four undershot waterwheels survive, of which three are 20 ft and one is 19 ft in diameter. These wheels drove two rows of stones, one of eight pairs and one of four pairs. The adjacent Miller's House was destroyed during the Blitz in the 1940s.The House Mill is Listed Grade I, and is now in the care of The River Lea Tidemill Trust, which has undertaken extensive refurbishment work, including replacing roof slates with tiles, replicating the Georgian facade of the Miller's House, providing a visitor reception area and exhibition space, and giving access to all floors of the mill.

The Clock Mill is the other principal building in this complex. It replaced an earlier timber building and survives largely as rebuilt in 1817. It is of similar size to the House Mill, being 80 ft long and of five storeys. The structure is of stock brick, with slate roof, and the distinctive architectural features include a weatherboarded lucam and an iron wall crane on the south elevation, and at the west end an elaborate clock tower and two conical kiln roofs. The interior structure is of timber beams and flooring supported on circular cast-iron columns The clock tower survives from the earlier mill, together with the clock, which is marked 'Charles Penton, Moorfields, 1753', and a bell founded in 1770 which hangs above in the open turret. Three iron undershot waterwheels survive in the wheel pit, one of 19 ft 6 in. and two of 20 ft diameter. The easterly wheel is marked on the rim with the maker's name; 'Fawcett & Co., Phoenix Foundry, Liverpool'. These wheels drove six pairs of stones at 130 rev./min. After the distillery closed the Clock Mill was converted into offices, but it is now disused.

2. Liberty's Printworks, Merton

HEW 2298
TQ 264 657

In the eighteenth century a flourishing south London textile industry was based on water power provided by the River Wandle. On the site of the Priory of Merton Abbey stood a calico factory, which was taken over in

Liberty's Printworks

DENIS SMITH

1885 by Arthur Liberty, of the *Regent Street* Store. Liberty used the mill for the dyeing of silks. Remaining on site is an undershot waterwheel 12 ft in diameter and 15 ft wide, of which the seven spokes and the rim were cast in one piece. Twenty-eight wooden floats are mounted on wooden 'starts' fitted in sockets in the cast-iron rim. Some gearing that drove the mill machinery survives. Also on site is the Listed Grade II eighteenth-century Colour House, which is built of brick with flint infilling.

Wimbledon Common Windmill

DENIS SMITH

3. Wimbledon Common Windmill

HEW 2299
TQ 230 725

Situated in *Windmill Road,* the earliest reference to a windmill on Wimbledon Common is for a post mill erected in 1613. The present mill is of an unusual type in Britain, namely, a hollow post mill. The vertical timber post supporting the weatherboarded structure is bored to take the iron drive shaft, with the wallower at the top, meshing with the brakewheel. The present mill was built in 1817 by Charles March, who was a carpenter rather than a millwright. The structure sits on an octagonal brick ground storey enclosing the machine floor. The next storey is of timber and encloses the stone floor. The roof of this storey includes the cross-trees supporting the hollow post and the quarter bars, which are enclosed to form a conical roof to the bin floor. The mill originally had patent sails and striking gear, and a fantail. In the 1860s the octagonal brick and timber machine and stone floors were converted to cottages by Lord Spencer, the owner. On the passing of the Commons Act in 1871 the mill came into the hands of the Wimbledon and Putney Commons Conservators, who inherited a building in poor condition. By 1890 it was in danger of collapse. A public appeal enabled a Lincolnshire millwright, John Saunderson, to undertake major reconstruction of the building, giving it its present appearance. Further work was done in the 1950s, but it was not until 1974, when the cottages were vacated, that the extent of damage by rot was revealed. After further restoration work the first floor was opened to the public in 1976 as a windmill museum.

4. Upminster Smock Mill

HEW 2303
TQ 556 869

This timber smock mill stands on high ground in *St. Mary's Lane* and was built in about 1800 by James Noakes. It is about 60 ft high and the patent shutter sails are 60 ft in diameter. The ground floor is enclosed by a brick wall, at the top of which is the projecting gallery, giving access to the sails. There are five floors: the dust floor under the cap, the bin floor, the stone floor, the meal floor and the ground floor. There are four pairs of stones. The machinery is not in operation. Noake's mill was working by 1803, and in 1812 a steam-driven mill was on site. A vertical, single-cylinder engine of 10 hp received steam from a 14 ft diameter Cornish boiler by Davy Paxman of Colchester. The steam mill was dismantled in the 1960s, and in 1965 the London Borough of Havering became the owner of the mill. The building is Listed Grade II and is open to the public free of charge at set times during the summer.

Manufacturing

5. Whitechapel Bell Foundry

HEW 2304
TQ 342 816

This foundry is one of the last of the hot-metal trades to survive in London. The firm was certainly established by 1570 when Robert Mot began making bells in Whitechapel. The foundry, which is situated in *Whitechapel Road*, was originally on the north side of the road, and

WENDIE TEPPETT

moved to its present site in 1738. The Georgian building, facing *Whitechapel Road*, is a three-storey, stock-brick structure behind which lies the foundry itself. The works comprise a cope and core moulding floor, the furnace and casting floor and departments for tuning (on a lathe) and wheel and frame making. The frames for hanging peals of bells are subjected to vibration and high stresses and were traditionally built in timber. The Whitechapel Foundry has actively promoted bell frames made in steel and, more recently, of reinforced concrete. Of the thousands of bells cast here the two most famous must be the Liberty Bell, cast in 1752, which survived an 11-week stormy Atlantic crossing to America, and Big Ben, cast in 1858, for the clock tower of the Houses of Parliament.

Whitechapel Bell Foundry

6. Royal Arsenal, Woolwich

HEW 2305
TQ 443 792

The Royal Arsenal, founded in 1671, together with the Naval Dockyard and the Royal Military Academy, made Woolwich an important centre for scientific, technical expertise and research during the nineteenth and early twentieth centuries. Woolwich Arsenal is the site of many

Royal Arsenal, Woolwich, Dial Square

DENIS SMITH

historic structures from the seventeenth, eighteenth and nineteenth centuries, 18 of which are listed buildings. Surviving buildings include the *Royal Laboratory* (1696, rebuilt 1802), the *Dial Square Arch Block* (1717), the *Royal Brass Foundry* (1717), the *Academy* building (1716–20) and the *Grand Store* (1806–13). Production of munitions expanded greatly during the nineteenth century, and by 1907 the site covered 1200 acres. At its peak over 80 000 people were employed, and many important engineers began their career here. During the twentieth century the Arsenal declined and it was closed as a Royal Ordnance Factory in 1967. Many of the buildings are in the care of the Ministry of Defence, which is actively seeking new uses for the buildings. The *Royal Laboratory* and the *Grand Store* are both Listed Grade II.

The former *Main Entrance* to the Royal Arsenal was built in 1829, but is now isolated from the rest of the Arsenal by road-widening works in the 1970s. It is Listed Grade II.

7. Whitbread Brewery, Chiswell Street

HEW 2307
TQ 325 819

In 1851 it was said that 'The Breweries of London may be considered as amongst its most important manufacturing establishments'. This was undoubtedly true in terms of the capital invested, the extent of brewery structures, their willingness to adopt new technology and, indeed, the age of the eight greatest porter breweries in the Capital.

Samuel Whitbread (1720–96) was born at Cardington in Bedfordshire and came to London in 1734 as an apprentice brewer. By 1742 he had established himself in business as a brewer. He moved to the present site in *Chiswell Street* in the City in 1749. The brewery buildings were developed over the years, and John Smeaton, James Watt and John Rennie were all associated with the work.

There was an important period in the 1780s. The Porter Tun Room was completed in 1782 and has a majestic 60 ft span king-post roof in timber with wrought-iron straps between the tie and the vertical posts. It is regarded as the largest king-post roof in Europe. The Tun Room was designed to house large slate vats holding 1½ million gallons of beer on a floor of 8400 sq. ft.

At the same time Samuel Whitbread conceived the idea of forming a stone-lined pair of vaults in which beer could be stored until racked off into barrels. Robert Mylne was consulted, but his vault leaked. John Smeaton was then consulted, and he produced a beer-tight design in York stone with joints pointed in tarras mortar. Smeaton invoiced Whitbread in June 1782 for vaults holding 9000 barrels for a design fee of £20 10s. Smeaton considered this a large sum for his design time and wrote:

> It looks like a momentous sum of money ... and yet I dare say you would not think one penny a barrel a great fee for design only; and this upon 9000 Barrels amounts to no less than £37 10.

Whitbread paid up in full and the two became good friends.

In June 1784, Boulton and Watt designed a 10 hp steam engine for Samuel Whitbread (only the second steam engine to be used in a brewery). This engine and its associated mill and pumping machinery was installed by John Rennie in 1785. The engine, which replaced 24 horses, was enlarged to 20 hp in 1814. Whitbread installed a new Boulton and Watt engine, of 30 hp, in 1841.

In 1783 Samuel Whitbread commissioned a portrait of John Smeaton by Thomas Gainsborough—The Institution of Civil Engineers has the copy by Wildman.

MATHIAS, P. *The Brewing Industry in England 1700–1830*. Cambridge University Press, Cambridge, 1959.

8. Kirkaldy Testing Works, Southwark Street

HEW 2155
TQ 318 812

David Kirkaldy (1820–97) established his independent firm for testing engineering materials in London, which served the construction, and other, industries from 1866 to 1965. Kirkaldy designed the unique testing machine himself and had it built, at his own expense, by Greenwood & Batley of Leeds. It is a universal machine 47 ft 6 in. long, and was designed to test in tension, compression, bending, torsion, shear, punching and bulging. The machine, capable of applying a load of 440 tons, was first installed in an existing building in *The Grove*, Southwark, and was open for business on 1 January 1866. Clients sent specimens for testing from all over the world. Kirkaldy was involved with the Institution of Civil Engineers' Steel Committee between 1866 and 1871 and with the Institution of Mechanical Engineers' tests on riveted joints. The present purpose-built building at *99 Southwark Street* was opened in January 1874. After the Tay Bridge disaster of December 1879, pieces of the wrecked bridge girders were recovered from the bed of the river Tay and brought to the Kirkaldy Testing Works for testing in the spring of 1880. The firm also did on-site structural testing,

DENIS SMITH

The Kirkaldy Testing Machine

notably of Wembley Stadium, which was built to accommodate 125 000 spectators. In 1923 the structure was tested for dead load (sandbags) and live loading, using hundreds of men who repeatedly stood, stamped and sat in unison. The test was devised and supervised by W. G. Kirkaldy. Three generations of the family ran the firm until the younger David retired in April 1965. The works, under new management, finally closed in 1974. The testing machine remains in position and in working condition as the centrepiece of The Kirkaldy Testing Museum.

SMITH, D. David Kirkaldy (1820–1897) and engineering materials testing. *Trans. Newcomen Soc.*, 1980–81, **52**, 49–65.

KIRKALDY, W. G. *David Kirkaldy's System of Mechanical Testing*. Sampson Low, London, 1891.

9. Ford Motor Company, Dagenham Works

HEW 2308
TQ 495 820

Henry Ford travelled to Britain to choose the site for his proposed new factory beside the Thames. The site he chose was the abandoned Corporation of London refuse site at Dagenham. Writing in 1930, Henry Ford said: 'We picked out the Dagenham site because it has water, rail and motor transport'. In total, 600 acres were purchased, of which 110 acres were used in the first instance. It was a difficult site in terms of foundation engineering, being on the Dagenham Levels that had been inundated several times by the infamous Dagenham Breach in the Thames bank (the remaining lake still lies within the works). Excavation began in May 1929. The factory is built on 22 000 concrete piles. It comprised a foundry (1260 ft by 300 ft), machine shop, and manufacturing and assembly shop (1000 ft by 300 ft), and included its own power station, with a 30 000 kW steam turbine supplied by three Babcock and Wilcox boilers. Forty-five coke ovens carbonising 800 tons of coal per day fed the single blast furnace. Coal and iron ore were delivered to a 1800 ft long jetty in the Thames capable of taking vessels of 12 000 tons. Storage areas capable of holding 34 200 tons of coal, 12 200 tons of iron ore and 11 500 tons of limestone were provided. An approach road, 1½ miles long, was built on embankments, together with 3½ miles of railway track. The architects were Charles Heathcote & Son of Manchester; the river jetty and power station were built by John Mowlem & Co. and Samuel Williams & Sons supplied the reinforced-concrete piles. The factory opened in 1931.

The Ford factory at Dagenham. *Concrete Constructional Engrng*, 1931, **26**, 8–19.

HARDING, SIR H. *Tunnelling History and My Own Involvement*. Golder, Toronto, 1981.

10. Trinity Buoy Wharf

HEW 2309
TQ 394 807

This important site is in *Orchard Place*, at the junction of the River Lea with the Thames at Bow Creek. Trinity House occupied it from 1803 until 1988 as a maintenance depot, and its 185-year tenure is an occupancy record for a Thames riverside site in Docklands. In 1822 the Elder Brethren of Trinity House consulted Ralph Walker to design and estimate the cost of rebuilding the original River Lea embankment in brick with a stone coping. His estimate was £1104 11s, and the work was undertaken by George Munday of Old Ford. This river wall survives and is the oldest structure at Trinity Buoy Wharf.

DENIS SMITH

Trinity Buoy Wharf

In 1851–52 the southern section of the River Lea wall and two-thirds of the Thames-side wall, including the river steps, were built to the designs of James Walker by Thomas Earle in high-quality ashlar stone at a cost of £4540. This work was listed as being of special architectural or historic interest in July 1983.

The largest of the nineteenth-century structures on site is the *Chain and Buoy Store* and the *Experimental Lighthouse,* which was designed by (Sir) James Douglass. The polygonal brick lighthouse rises against the east wall of the stores to a stone cornice supporting a railed gallery at the base of the lantern. The lantern of 1866 was made by

Campbell, Johnstone and Company. These buildings were constructed by J. F. Stewart at a cost of £4065. It was in this lighthouse that Michael Faraday carried out the first experiments in electric lighting for lighthouses.

In 1869 the Corporation of Trinity House established an engineering workshop to repair and test the new wrought-iron buoys then coming into use. In May 1875 plans for a chain and cable proving-house were drawn up by James Douglass and the building was completed by the end of the year. In 1952–53 the new *Fitting Shop*, with a corrugated concrete shell roof, was built.

Trinity House closed Trinity Buoy Wharf on 31 December 1988, the work was transferred to Harwich, and the site was sold to the London Docklands Development Corporation.

Civil Aviation

11. Hendon Aerodrome

HEW 2090
TQ 222 903

Hendon Aerodrome played an important role in the early days of flying. Claude Grahame-White established an aircraft factory and flying school here. The Grahame-White Company offices (built 1915) in Aerodrome Road survive and are Listed Grade II. The venture was flourishing when, in 1914, the Government requisitioned the aerodrome, which remained a Royal Air Force base until 1968.

Part of the site is now occupied by the RAF Museum and two of the First World War hangars survive in use as exhibition galleries. These hangars date from 1917–18 and are notable examples of the use of Belfast roof trusses. This type of roof truss, of timber lattice construction, was in common use at that time to provide moderately large span roofs. The roof trusses are of bowstring form, with a rise of about 10 ft. The hangars at the museum are each made up of two bays of 80 ft span and are about 171 ft long.

At the time of writing, a scheme is in hand to re-erect the Grahame-White hangar, built in 1915–16, which stands on the adjacent Ministry of Defence site, within the museum boundary, where it will be used to display First World War aircraft. This structure is in two sections and seems likely to have been built and used as a workshop. The earlier part, believed to date from 1917, has a Belfast roof truss of about 100 ft span. Steel trusses carry the roof of the extension, built in 1918.

FRANCIS, P. *British Military Airfield Architecture*. Patrick Stephens, Yeovil, 1996.

12. Croydon Airport

HEW 2310
TQ 312 636

London's earliest civil airport, with customs facilities, was based at Hounslow Heath, and it was from here that the first scheduled London–Paris flight took place on 25 August 1919, carrying just one passenger. By March 1920 Hounslow Heath ceased to be used for passenger traffic and Croydon became the Capital's air terminal. Croydon was only 10 miles from central London, and much of the inter-War development in civil aviation took place here. Prior to the First World War the site had been *New Barn Farm*. During the War 'temporary' timber buildings were erected in 1915 for the Royal Flying Corps, and from 1920 to 1927 the airline companies had to make use of these buildings. However, access to the airport benefited from the construction, in 1925, of *Purley Way*—part of the Croydon Bypass. In 1926 work began on the replacement of the First World War buildings with London's first purpose-built international airport accommodation. The buildings, facing *Purley Way*, were designed by the Air Ministry's *Directorate of Works and Buildings* and comprised a control tower and a passenger terminal. The 50 ft control tower is placed centrally to the passenger building; both are steel-framed buildings clad in concrete block-work. The control tower has a cantilevered balcony between the second and third storeys. The two-storey passenger building was the prototype of subsequent airport terminal planning, with a spacious hall surrounded by offices, beyond which were passport controls, customs hall, and departure gate. The terminal buildings are now known as *Airport House*, and alongside is the *Aerodrome Hotel*. The new airport facilities were opened in 1928. In March 1924 the Government amalgamated four airlines to form *Imperial Airways*, which based themselves at Croydon.

However, Croydon Airport, with its undulating grass field and severe shortage of hangar accommodation, was

Croydon Airport, control tower

DENIS SMITH

already inadequate as London's main airport even before the Second World War. *The Sphere* journal of 5 February 1938 said 'In fact, one of the air lines, British Airways, merely uses Croydon as a flying base and flies its machines to Gatwick every night for servicing and back to Croydon in the morning before starting the day's work'. After the Second World War Croydon was perceived as unsuitable for further development, as Gatwick, which had opened in 1936, had developed as the principal south London airport.

13. Heathrow Airport

HEW 2311
TQ 080 755

After the War the Government decided that from 31 May 1946 the London Airport should be at Heathrow, a former RAF airfield 14 miles from central London. Initially, the existing RAF runways were used, and there were no terminal, baggage-handling or administration buildings. Customs inspection and immigration control all took place in marquees and caravans. In the first 12-year phase of the development, Wimpey's were the main contractors working on a £13 million scheme. The first permanent buildings opened in 1955 and included a control tower, the Queen's Building (offices and restaurant) and a terminal for short-haul traffic. The new facilities were connected to the *Bath Road* by a road tunnel, which is still in use. British European Airways had a hangar building with slender prestressed concrete roof beams of 150 ft clear span. The building comprised ten hangars, each 110 ft by 180 ft, and was designed by (Sir) Alan Harris and completed in 1952. This important structure has recently been demolished.

The BOAC Maintenance Building (TQ 092 756) is a large-span reinforced-concrete building which was designed by Sir E. Owen Williams for The Ministry of Civil Aviation and leased to British Overseas Airways Corporation on completion. Work on site began in September 1950. The structure is of reinforced concrete and provides nearly 1 million sq. ft of hangar, workshop and office accommodation. The rectangular plan comprises four large hangars 140 ft deep, with unprecedented clear openings each of 336 ft with motorised folding doors. To achieve this clear opening two 1000 ton counterbalances support a pair of 90 ft cantilevers spanned by a central inverted V-beam. A cruciform spine structure containing offices and maintenance workshops separates the four hangars. The overall flat roof is formed of ridge-and-furrow glazing. This building is undoubtedly Williams' greatest achievement in planning and structural design using *in situ* concrete. The contractors were W. & C. French.

Sir Owen Williams also designed the BOAC 'Wing Hangars' built in 1954–56. The building comprises two hangars, each 110 ft deep and 565 ft long, flanking the

60 ft wide central workshop building, giving the building a ground plan of 565 ft by 280 ft. The roofs of the hangars are cantilevered from the two-storey reinforced-concrete frame of the workshop building and supported by inclined ties of mild steel at three points along the length of the hangar. The 'Wing Hangars' are so called because only the front ends of the aircraft are covered by the hangar while the tail end projects outside the enclosure during routine maintenance. The client was The Ministry of Civil Aviation and the contractors were again W. & C. French.

COTTAM, D. *Sir Owen Williams: 1890–1969*. Architectural Association, London, 1986, 129–37.

HARRIS, A. J. Hangars at London Airport: design of large span pre-stressed concrete beams. *Struct. Engr*, October 1952, 226–35.

Communication Industry

14. Enderby's Wharf, Greenwich

HEW 2313
TQ 393 786

This site on the west side of the Greenwich peninsula is of prime importance in the history of the communications industry in London. The Enderby family, who had been engaged in the whaling industry and in exploration, established a ropewalk on this site in 1834. The first successful transatlantic telegraph cable was made here, and was laid by Brunel's *Great Eastern* in 1866. The 5000 miles of cable were insulated at the Gutta Percha Company's works at *19 Wharf Road*, adjacent to the City Road Basin of the Regent's Canal. The cable was then brought to the Telephone Construction and Maintenance Company's works at Enderby's Wharf in Greenwich. The wire-covered cable was paid out from the wharf into tanks in a hulk lying in the Thames. The hulk was then towed by steam tug down to the *Great Eastern* lying off Sheerness. The *Great Eastern* then sailed to Valentia in south-west Ireland, where the British end of the cable was to be located. On 23 July 1865 the ship left Ireland, but after laying over 1000 miles of cable the cable broke and the attempt was abandoned. On 13 July 1866 the *Great Eastern* again sailed from Valentia, and on the 27 July the cable was successfully completed to Trinity Bay Newfoundland. This impressive technical feat led to enormous investment in future cables, and the Enderby Wharf works in London played a major role in the expansion of communication throughout the world.

15. Post Office (British Telecom) Tower

HEW 246
TQ 292 819

This structure was originally designated *The Museum Radio Tower, London*, as the site chosen for its erection was adjacent to the existing museum telephone exchange in Howland Street. The tower project evolved in response to developments in UHF, or microwave, radio-telephone and television communications. The design brief required line-of-sight communication with other towers, and the growth of high buildings in London necessitated a tall structure. The project also required firm foundations and a stiff tower structure to maintain aerial alignment. The design comprised a 620 ft high tower with a hollow central shaft with circular cantilevered floors. With rigidity a prime consideration, a 1:67 scale model, 8 ft 2 in. high, was tested in the wind tunnel at the National Physical Laboratory in 1961, with results that confirmed the structural analysis.

The foundation was required to carry a load of 13 000 tons, and the design adopted was for a 90 ft square prestressed concrete base on the blue clay 22 ft below street level. The circular shaft was slip-formed to reduce high-level scaffolding and the core was 35 ft in diameter up to the 205 ft level, and above that level 24 ft in diameter. From the 115 ft to the 355 ft levels the tower is clad in glazed panels and has an overall diameter of 52 ft. The microwave aerials are placed between the 355 ft and 477 ft levels. From 477 ft to 550 ft the space was dedicated to public access, including a revolving restaurant floor of 64 ft diameter, but the public are no longer allowed into the tower. Above this level are the lift motors and tanks.

Construction began in July 1961 and the main structure was completed in July 1964. The client for the tower was the General Post Office, and the design and

Post Office
(British Telecom)
Tower

WENDIE TEPPETT

construction work was by the Ministry of Public Buildings and Works, for whom the main contractors were Peter Lind & Co.

CREASY, L. R., ADAMS, H. C. and LAMPITT, N. Museum Radio Tower, London. *Proc. ICE*, 1965, **30**, 33–78.

Entertainment and Leisure

16. British Museum and Reading Room Roof

HEW 304
TQ 301 817

The British Museum was established by Act of Parliament in 1753 to bring together items in various collections. The seventeenth-century Montagu House, standing on the present site of the museum was purchased and opened in 1759 as The British Museum. It stood until 1845 but the present King's Library, the first new building on the site, was completed in 1826, and the new south wing with the present entrance portico was completed in 1847. The King's Library is notable for the large cast-iron beams spanning 41 ft, with open webs to reduce the weight and integral sockets for secondary timber beams. Foster Rastrick & Co. supplied the iron beams in 1824–25.

The magnificent roof over the Reading Room fills the courtyard, which was previously in the centre of the building. The roof comprises a dome 140 ft in diameter, formed of 20 cast-iron ribs springing from the base and united at the top by a circular ring surmounted by a lantern 40 ft in diameter. The main ribs are 106 ft high, and brick arches span between the ribs above the glazed area. The dome is supported on 20 cast-iron columns encased in concrete; the circular wall between the columns is non-structural. The form of the roof was suggested by (Sir) Anthony Panizzi, then principal librarian, and Sidney Smirke, the Architect to the Trustees, designed the details. The builders were Baker and Fielder and the ventilation system was designed and installed by Haden and Son of Trowbridge. The work was completed in 1857 at a total cost of £150 000. The dome was damaged by a bomb in 1940, but an engineering examination of the structure in 1963 found the ironwork perfectly sound.

The space between the circular Reading Room and the surrounding rectangular courtyard has been converted into a magnificent glazed, covered piazza, providing a new circulating space known as the *Great Court*. Work began on site in March 1998 with demolition of the book stacks in the space surrounding the reading room. The design of the geodetic steel roof lattice of the Great Court was complicated by the fact the centre of the reading room was 5 m north of the intersecting diagonals of the courtyard. The roof comprises 5200 members, which had to be welded to 1800 steel nodes to carry 3312 glass panels. The roof weighs just under 800 tonnes: 478 tonnes of steel and 315 tonnes of glass. The newly created piazza was opened to the public in 2000. The work was funded by Millennium Commission and Heritage Lottery grants. The engineer was Buro Happold and the architects were the Foster Partnership.

ANDERSON, R. *The Great Court and the British Museum*. The British Museum Press, London, 2000.

17. Palm House, Royal Botanic Gardens, Kew

HEW 366
TQ 187 769

The Royal Botanic Gardens consist mainly of two estates acquired by the Royal Family during the eighteenth century. The small botanic garden was opened to the public in 1841 and has expanded to its present size of more than 121 acres. There are several interesting buildings on the site, but the most important from a structural point of view is the *Palm House*, with its skeletal structure of rolled wrought-iron ribs. The design had its origins in 1843, when Richard Turner of the Hammersmith Ironworks in Dublin visited London in connection with a winter garden in Regent's Park. Turner met the botanist (Sir) William J. Hooker, Director of the Royal Botanic Gardens, and in January 1844 Turner impressed Hooker with plans for a palm house. The arched nave shape of the building had been determined in outline by the architect Decimus Burton. At this stage the roof arch ribs were to be of cast iron, and it was Turner who suggested replacing them with wrought iron. In 1844 Kennedy & Vernon had produced their patent for rolled wrought-iron I-beams for ship-deck beams, and their use in the Palm House was to be the first in an architectural building. The building has an overall length of 362 ft and the iron arch ribs are placed at 12 ft 6 in. centres. The wrought-iron ribs were rolled at Malins & Rawlinson's works at Millwall under licence from Kennedy & Vernon of Liverpool. The 9 in. deep straight bars were shipped from Millwall to Turner's Works in Dublin, where they were joined and rolled to the correct curve. They were then shipped back to Kew, where the first rib was erected on 15 October 1845. The innovative building was completed in 1848.

18. Crystal Palace

HEW 305

Hyde Park: TQ 273 798

Sydenham: TQ 339 711

The Great Exhibition held in Hyde Park in 1851 ran from 1 May to 15 October, attracted 6 039 195 visitors and resulted in a surplus of £186 437, which was used to purchase and develop the South Kensington estate, including the Royal Albert Hall, the museums and Imperial College. The Exhibition site occupied 26 acres between Rotten Row and South Carriage Drive. The building, of cast iron, glass and timber, was designed on a 24 ft square grid, with a central, north–south transept and a timber arched roof of 72 ft span. The glasshouse concept of Joseph Paxton was realised by William Cubitt, and the ironwork contractors were Fox Henderson & Co. William Cubitt, President of the Institution of Civil Engineers, was Chairman of the Building Committee and was knighted at the end of the Exhibition. In 1852 the Crystal Palace Company was formed, with George Grove, a civil engineer, as Company Secretary. The Company bought the Hyde Park building from the Commissioners. The modular building was eminently demountable, and their first task was to remove the building from Hyde Park and to use the materials as part of a very different structure at Sydenham.

Construction of the Palace on Sydenham Hill began in the early autumn of 1852, and the building was opened by Queen Victoria on 10 June 1854. A comparison of the two structures is instructive (Table 5).

At the north and south ends of the Sydenham building water towers 282 ft high, designed by I. K. Brunel, carried tanks that were used for the fountains on the terraces. The Crystal Palace at Sydenham was destroyed by fire on the night of 30 November 1936. In the early 1940s Brunel's water towers were demolished as they were considered landmarks for enemy bombers. Rail access to the site was by two railway stations: the High and the Low Level. The *High Level Station* (1865) fronted Crystal Palace Parade and all that now remains are some arcaded

Table 5: Comparison of the Crystal Palace structures at Hyde Park and Sydenham

	Hyde Park	Sydenham
Length (ft)	1 848	1 608
General width (ft)	408	312
Maximum width (ft)	456	384
Height of nave (ft)	63	104
Height of transept (ft)	108	168
Ground floor area (sq. ft)	772 784	598 396
Area of galleries (sq. ft)	217 000	245 260
Total floor area (sq. ft)	989 884	843 656
Area of glass (sq. ft)	900 000	1 650 000

walls, but the pedestrian subway comprising three rows of octagonal columns in brick with flared capitals remains, although it is not easily accessible. The *Low Level Station* was built by the London, Brighton & South Coast Railway in 1854. It was modified in 1875 and refurbished in 1979. The remains of the station are Listed Grade II. The Independent Television transmission mast was built on the site of Brunel's north water tower.

19. The German Gymnasium

HEW 2314
TQ 301 831

The roof of the former German Gymnasium, in *St. Pancras Road* close to King's Cross Station, is supported by laminated timber arches, a structural form often used in the early nineteenth century as an alternative to trusses in timber or iron.

In the late eighteenth and early nineteenth century a gymnastics movement became very popular in Germany, and a number of clubs were set up by Gymnastic Societies. In 1861 the German community in London formed the Germans Gymnastic Society and a sum of over £30 000 was rapidly raised for a club building, which included a library and other facilities as well as the main exercise hall. The hall required a wide space uninterrupted by columns, and to achieve this the architect, E. Gruning, designed arch ribs built up of horizontally laid planks, or laminations, of timber bolted together and curved to form an arch acting in combination with a more conventional pair of beams or rafters which supported the purlins of the roof.

Laminated timber had been developed as a means of producing substantial lengths of wood that were not available from 'nature'. The use of horizontally laminated arches was pioneered by the Bavarian engineer Carl Wiebeking in 1805–10 for bridges, and the French engineer Armand Amy, *c.* 1818–40, for roof structures. Wiebeking's work had been publicised in England by Thomas Tredgold, and from about 1827 the Newcastle-based engineer and architect John Green, later joined by his son Benjamin, developed similar designs, initially for bridges, but by the 1840s for roofs. It was seen as an economic alternative to both iron and masonry. By the mid-1840s the technique was well known through the publications of these pioneers. The Greens' work on Tyneside railways was used by Joseph Locke for bridges in France and by Lewis Cubitt at King's Cross Station. Laminated timber roofs were also being used in textile mill buildings of the time, so there were precedents for Gruning's work at the Gymnasium. As most earlier examples have now disappeared, however, the gymnasium is of considerable interest, and its proximity to King's Cross Station enables one to obtain a good

The German Gymnasium

impression of how the original roof there would have appeared.

The Gymnasium hall is 129 ft long, 80 ft wide and 57 ft high. Work began in 1864 and the hall was opened, although not completed, in 1865. The contractors were Piper and Wheeler and the cost was approximately £6000.

Another surviving example of a laminated timber roof is at St. Paul's Presbyterian Church (1859), West Ferry Road, Isle of Dogs (TQ 372 789).

YEOMANS, D. T. (ed.). *Timber as a Structural Material*. Ashgate, Aldershot, 1999.

The Builder, 19 May 1866, 366–67.

20. Bethnal Green Museum

HEW 2315
TQ 351 829

The Museum, in *Cambridge Heath Road*, is a branch of the Victoria and Albert Museum, and was opened on this site in June 1872 by HRH the Prince and Princess of Wales. The Museum's iron structure was part of the temporary buildings erected in 1856 for the South Kensington Museum, the predecessor of the present Victoria and Albert Museum.

The building has three bays with wrought-iron arched and trussed roofs, each of 42 ft span, supported on cast-iron columns 26 ft high. The central nave is the full height of the building and the side aisles have mezzanine floors with a grid of wrought-iron beams at 7 ft centres. The hollow cast-iron columns were used to drain the valley gutters and to provide ducts for gas pipes for lighting the interior. The building was designed by Major-General H. Y. D. Scott of the Royal Engineers. Scott also designed the Royal Albert Hall.

The building now houses the Victoria and Albert's Museum of Childhood.

21. Royal Albert Hall

HEW 300
TQ 266 796

The South Kensington site was developed by the Commissioners of the 1851 Exhibition funded by the surplus from the Crystal Palace. Many new buildings for various institutions were built. The first was the South Kensington Museum (1856), the prototype Victoria and Albert Museum, and the second was the Albert Hall. It

Royal Albert Hall under construction

was built in 1867–71 to the overall design of H. Y. D. Scott, RE, to fulfil two functions, namely a concert hall and a conference centre for learned societies. The predominant engineering feature of the hall is its magnificent wrought-iron roof. The roof design presented special design problems. Two consulting civil engineers, J. W. Grover and R. M. Ordish, made the detailed design and calculations. An overall Advisory Committee had been appointed in April 1866, including architects, artistic designers and two civil engineers, John Fowler and John Hawkshaw, who both made suggestions during the design process. In addition, William Fairbairn of Manchester, the ironwork contractor for the roof, also made significant contributions to the details of the design.

The roof is an elliptical trussed dome 219 ft 4 in. by 185 ft 4 in. Thirty radial trusses deliver their load onto a horizontal-web plate girder sitting on top of the masonry wall 120 ft above street level. The central ring curb is also elliptical and is 17 ft 6 in. deep, the maximum depth of the radial trusses. A trial assembly was made at

Fairbairn's Engineering Works, dismantled, and sent to London, where it was erected from a huge temporary timber centring scaffolded from the arena floor and completely filling the auditorium. In May 1870 the wedges were knocked out from the centring and the iron roof structure settled 5/16 in. The roof has ridge-and-furrow glazing to admit daylight, and beneath the glass the glare was reduced by a draped textile *velarium*. The hall was opened by Queen Victoria on 29 March 1871 and named to commemorate Prince Albert who had died in 1861.

Electric lighting was installed in 1879. In 1941 the first real attempt to correct the poor acoustic characteristics was made when a sound-reflecting canopy over the platform was suspended from the roof. In 1949 the velarium and the glazed inner dome were removed and replaced by the present fluted aluminium double-skin dome fixed to the bottom ribs of the roof trusses. During 1968–69, 109 fibreglass acoustic diffusers were also suspended from the roof in a further attempt to improve the acoustics.

Illustrated London News, 8 April 1871, 346.

CLARK, R. *The Royal Albert Hall*. Hamish Hamilton, London, 1958.

22. South Kensington Subway

HEW 2316
TQ 269 787 to
TQ 268 793

South Kensington Station on the Metropolitan District Railway was opened in 1868. From there it was an uphill approach to the Royal Albert Hall, and it was considered that a railway in a subway under *Exhibition Road* would be useful. In 1872 an Act (35 & 36 Vict. c.192) was obtained to build a subway railway, to be operated on the pneumatic principle, from South Kensington Station to the Albert Hall, but the finance was not forthcoming. It was not until 1885 that the Metropolitan District Railway opened their pedestrian subway, which remains in use today. Originally there was a toll of 1d until it was freed from tolls in 1908. The subway has vertical side walls that are spanned by a brick barrel vault roof as it passes under *Cromwell Road* and elsewhere by transverse riveted wrought-iron girders with brick jack-arches springing from the bottom flange. Glazed ventilators and light-wells, with cast-iron framing, are visible at street level.

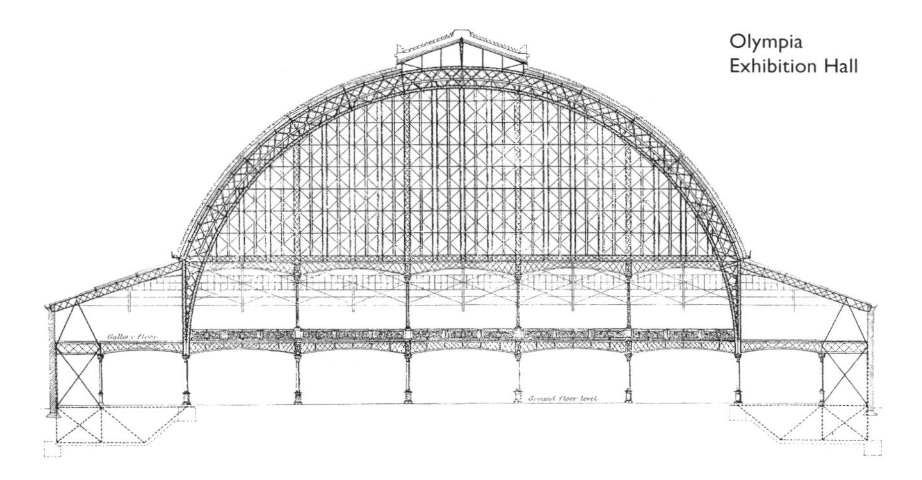

Olympia Exhibition Hall

23. Olympia Exhibition Hall

HEW 773
TQ 243 790

This building, with its dramatic iron arched roof, was promoted by the National Agricultural Hall Company and stands on a site of 6¼ acres, of which the Great Hall covers 2½ acres. It was opened in 1886. There is a clear 440 ft long exhibition space spanned by an iron trussed arch roof spanning 170 ft and rising to 99 ft 7½ in. above floor level. The arch ribs are 7 ft deep and are spaced at 34 ft centres. The inner portion of the gallery is carried on columns with a ball-and-socket joint top and bottom. Kensington (Olympia) Station, adjacent to the hall, is served by a number of main-line trains and by a London Underground shuttle service operating from High Street Kensington and Earls Court. Henry E. Coe, the company's architect, designed the hall, and the joint engineers were Arthur T. Walmisley and Max van Ende. The general contractors were Lucas & Son and the ironwork contractors were Handyside of Derby. The *Hammersmith Road* facade was rebuilt in 1929–30.

WALMISLEY, A. T. *Iron Roofs*. E. & F. N. Spon, London, 1888, 71–76.

24. British Empire Exhibition Site, Wembley

HEW 2317
TQ 193 857

The Exhibition was opened on 23 April 1924 and closed in November the same year. It was re-opened in May 1925 and finally closed in October of that year. The Exhibition led to the construction of 15 miles of new roads, a new Metropolitan Railway station (Wembley Park), and impressive exhibition buildings in reinforced concrete on a 216-acre site. The buildings were designed by the engineer (Sir) E. Owen Williams and the architects Simpson & Ayrton. The buildings included the *Palace of Engineering* (demolished in the 1970s), the surviving *Palace of Industry*, the facade of the *Palace of Art* (Listed Grade II) and the *Wembley Stadium*. The Palace of Industry is still in use and comprises a series of halls, each about 50 ft wide with pitched roofs. The columns and the knees of the portal frames are cast *in situ* with precast open-web rafters. Future uses for the facade of the Palace of Art are being discussed, and include its incorporation into a hotel building. Part of the elevated reinforced-concrete track of the screw-driven 'Neverstop' railway survives adjacent to Wembley Park Station. The railway was used to transport visitors within the Exhibition site. E. Owen Williams received a knighthood after the closure of the Exhibition.

Subsequent to the closure of the exhibition the owners of the site chose Sir Owen Williams to design the *Empire Pool*, which was built in 1933–34. The building was designed to house one of the largest swimming pools in the world (200 ft by 60 ft) and was built within the original exhibition grounds on the site of the large artificial lake. From the outset the building was required to provide a roof over the pool (which was to be readily converted into an ice rink or other display arena) and to cover the tiered seating for 4000 spectators. Williams' solution was to provide *in situ* concrete frames having a clear span of 236 ft. Externally each of the frames is exposed and they appear as a series of fins. The building is now known as the *Wembley Arena* and hosts a wide variety of activities.

COTTAM, D. *Sir Owen Williams: 1890–1969*. Architectural Association, London, 1986.

25. Royal Agricultural Hall, Islington

HEW 562
TQ 314 836

This exhibition hall lies between *Liverpool Road* and *Upper Street*, Islington, with the main entrance in Liverpool Road. It was opened in 1862. The architect was Frederick Peck, the building contractors were Hill, Keddell and

Robinson, and the ironwork contractors were Handyside of Derby. The hall was built to house the Smithfield Show and the foundation stone was laid by Lord Berners on 9 November 1861. The *Main Hall* (217 ft by 384 ft) is spanned by iron latticed web arch ribs of 130 ft span sitting on cast-iron columns at 24 ft centres. The hollow columns were also used to drain the roof area. The *Gilbey Hall* (151 ft by 233 ft) was added in 1894 and the *Prince's Gallery* was completed in 1920. The hall has hosted a variety of events, such as bakery, brewery and cycle exhibitions, the first Motor Show, the Dairy Show, dog shows, military tournaments, circuses and the World's Fair. In 1943 the Post Office Parcels Office at Mount Pleasant was burnt out and the department moved to the Royal Agricultural Hall. The Post Office left in 1971 and the building remained empty until 1976 when the London Borough of Islington bought the property. It is now a Business Design Centre.

The Builder, 21 September 1861, 653.

26. Alexandra Palace, Muswell Hill

HEW 2318
TQ 296 901

The success of the Crystal Palace in Sydenham, south London, inspired the idea for a similar centre on Muswell Hill in north London. In 1863 the Alexandra Park Company was formed to acquire land as a site for public recreation. The company negotiated with Kelk and Lucas to buy part of the structure of the 1862 Exhibition at South Kensington. They paid £80 000 for the materials and spent a further £90 000 on foundations at Muswell Hill. Work began in June 1864 to designs by Alfred Meeson and his partner J. Johnson, but the work was delayed by shortage of money and soil subsidence. In 1865 the Alexandra Park Company went into liquidation, and four subsequent companies suffered a similar fate. Eventually, after nine years, the Palace was opened on 24 May 1873, having cost a total of £504 724. Just 16 days later, on 9 June, it was destroyed by fire. It was decided to rebuild, and Meeson and Johnson were commissioned to prepare new designs. Work began on 10 October 1873 and was completed in July 1874 at a cost of £417 128. The Palace was opened on 1 May 1875 by the Lord Mayor of London. The building was larger than the first Palace, with a central transept

DENIS SMITH

Alexandra Palace

386 ft long and 184 ft wide. Two lines of cast-iron columns, surmounted by semi-circular wrought-iron arches of 85 ft span form the Great Hall. The 1354 tons of cast ironwork was supplied by the Staveley Ironworks and the 764 tons of wrought iron were supplied by Handyside of Derby. Continuing financial difficulties led to an Act of 1900 that enabled Middlesex County Council and local district councils to purchase the Palace and to manage its affairs. On 2 November 1936 the BBC inaugurated the world's first regular high-definition television service here, providing a welcome source of income. In 1966 the Greater London Council became responsible for the Palace. In September 1980 the ill-fated building was severely damaged by fire and was again restored. The Palace is Listed Grade II.

The New Alexandra Palace. *The Builder*, 15 August 1874, 687.

27. Ritz Hotel, Piccadilly

HEW 2334
TQ 291 803

The provision of hotel accommodation in the Capital provided the opportunity for structural innovation by engineers and ingenuity by contractors. Hotel construction sites often presented difficult access and site storage problems for the contractors. These difficulties, and

WENDIE TEPPETT

Ritz Hotel, Piccadilly

associated traffic problems, led to the need for rapid construction and new materials and techniques. Before 1900 the steel skeletal fame had been used in the United States in the construction of skyscrapers, but the earliest use of the full structural steel frame in an important building in London was for the Ritz Hotel in *Piccadilly* built in 1904. Nevertheless, the London Building Act of 1894 still required the external walls to be of full load-bearing thickness, denying the designers the full advantages of steel-frame construction.

The widespread publicity given to the construction of the Ritz Hotel, Selfridges' extension and other Edwardian steel-framed buildings led the London County Council to introduce the Building Act of 1909 which, belatedly, sanctioned steel-frame construction.

ENGLISH HERITAGE. *Early Structural Steel in London Buildings*. English Heritage, October 2000, Survey Report.

BYLANDER, S. Steelwork in buildings—thirty years' progress. *Structural Engineer*, January 1937, 2–25.

The Dorchester Hotel

WENDIE TEPPETT

28. Dorchester Hotel

HEW 2335
TQ 283 804

This hotel in *Park Lane* attracted a great deal of interest during its construction, largely because of the speed of its construction. It was an early example of a building where the structural advantages of reinforced concrete were fully employed. The project was also distinguished by interesting management techniques. At the beginning of the work changes were made in both the design and organisation, resulting in the engineers and architects finally responsible for the work being appointed after certain foundation and preliminary work had been done. At the 20 April 1930 the site was little more than a hole in the ground, yet the hotel was opened exactly a year later on 20 April 1931. The project comprised a 400-room building at a cost of somewhat over £1 700 000.

The speed of construction resulted from a variety of factors, including close co-operation between architects, structural and services engineers and contractors. Design decisions were also important, including: bedroom floor planning made independent of lower construction by heavy slab construction at first-floor level; elimination of columns above first-floor level by use of structural walls; and the use of high-quality concrete to minimise the size of elements. The external walls are 7 in. thick with 4 in. thick precast cladding panels made of concrete with a marble aggregate. The inner face of the external walls was lined with cork panels to provide heat and sound insulation. Above the first-floor slab the structural work progressed at the rate of one floor per week. Between 30 and 40 architects, engineers and surveyors were employed on the work, apart from outside staff.

The architects were W. Curtis Green and Partners, the consulting engineers were Considere Constructions Ltd. and the contractors were Sir Robert McAlpine & Sons.

Concrete and Constructional Engineering, 1931, **26**, 288–303.

29. Commonwealth Institute, Holland Park

HEW 2336
TQ 250 794

The first Commonwealth Institute building was demolished to allow extensions to the Imperial College of Science and Technology. The new Institute was built on the

The Commonwealth Institute

WENDIE TEPPETT

edge of Holland Park about a mile to the west of the old site. The building was designed to accommodate an exhibition space, a cinema, an art gallery, a restaurant and offices. The building was opened in the autumn of 1962, and its principal distinctive feature is the tent-like roof to the exhibition area.

In 1960 a site survey revealed that piling was not necessary, and the foundations are spread footings with a maximum ground pressure of 2 tonf/sq. ft. The contractor took possession of the site in October 1960 and concreting of the basement floor began in January 1961.

The 155 ft square ground floor of beam and slab construction is of 12 in. thick reinforced concrete with 18 in. diameter columns on a square grid of 26 ft 9in.The first- and second-floor galleries are also 12 in. thick and are connected by steps or ramps. The central paraboloid roof is 93 ft square, of prestressed concrete that is generally 3 in. thick but increases to 7 in. at the junctions with the prestressed triangulated edge beams. The central paraboloid shell has a geometric grid of rectangular generators and, surrounding this, four warped surfaces (with radial ribs and a broad eaves overhang) extending the plan of the roof to a square of 183 ft sides. Each rib in the outer warp is of a different length and has differently twisted top surfaces. Freysinnet prestressing was used extensively. The roof is covered with copper on an insulating layer of 1¼ in. vermiculite. The Institute was opened by Her Majesty the Queen on 6 November 1962.

The consulting engineers were A. J. & J. D. Harris, the architects were Robert Matthew, Johnson-Marshall & Partners and the main contractors were John Laing Construction Ltd.

SUTHERLAND, R. J. M. and POULTON, V. T. The Commonwealth Institute. *The Consulting Engineer*, May 1962, 500–03.

30. Millennium Dome

HEW 2320
TQ 391 802

The architects for this structure on the Greenwich peninsula were the Richard Rogers Partnership and the consulting structural and services engineers were Buro Happold. Construction was managed by the McAlpine–Laing joint venture. The structure comprises a network of cable netting suspended from 12 steel masts and covered by a dome of PTFE-coated glass fibre. The construction was managed by a consortium combining Sir Robert McAlpine and John Laing Construction. Construction began on 23 June 1997 when the first of the 8000 concrete piles was driven by Keller Ground Engineering of Coventry. The foundation work on site included drainage, service trenches, the ground-floor slab, sewage disposal, bases for the 12 masts and the reinforced-concrete ring beam, which is 6 m wide and 0.5 m deep and takes all horizontal and vertical forces from the cables. All this foundation work was undertaken by John Doyle

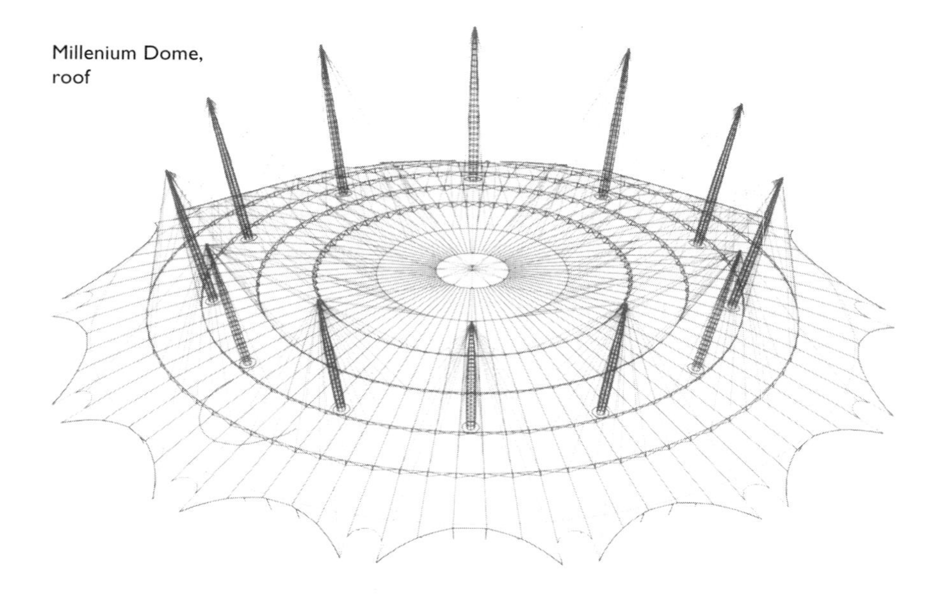

Millenium Dome, roof

Millenium Dome, erection of masts

Construction Ltd. of Welwyn Garden City. Watson's Steel Ltd. manufactured the twelve 90 m long masts which were erected on site in October 1997. Watson's contract also included the 70 km of high-strength steel cable. During the last two weeks of October 1997 the cable netting took shape on the ground and within a few weeks was winched up into position. On 23 March 1998 the first fabric roof panels, made from Teflon-coated glass-fibre fabric by Chemfab in New Hampshire, U.S.A., were hoisted. The 100 000 sq. m of fabric were installed by Birdair (Buffalo, U.S.A.) in 144 panels comprising 72 segments of the dome. The topping-out ceremony, celebrating the completion of the Dome structure, occurred on 22 June 1998. The Dome is served by *North Greenwich Station* on the Jubilee Line Extension Railway. The structure is an open-box, cut-and-cover structure, 405 m long by 32 m wide. Its construction required the removal of 300 000 cu. m of material and is the largest underground station in the world.

WILHIDE, E. *The Millennium Dome*. Harper Collins, London, 1999.

31. British Airways London Eye

HEW 2361
TQ 307 799

This dramatic structure had its origins in a design competition for a central London Millennium project—a competition that nobody won! However, the architectural design concept was pursued by David Marks and Julia Barfield, with analytical structural input from Ove Arup and Partners. Wind loading on the structure was thoroughly assessed and both the radial and circumferential cables were prestressed. Once the design had matured in London the scheme became a multi-national constructional project. The structural steel rim of the wheel was fabricated by Hollandia in The Netherlands, the passenger capsules were built in France and clad in double-curvature laminated glass from Italy. One of the design features was that the wheel should project over the river. Piled dolphins were constructed in the river to receive the prefabricated quadrants delivered by sea and river to be assembled in the horizontal position. The wheel was lifted into the vertical position on 10 October 1999 and the last capsule was fitted on 11 November. Although the fitting of the first capsule took four hours this process was soon reduced to 90 minutes.

Each of the 32 capsules is maintained in the horizontal position by a computer system and twin tilt switches linked to an incremometer that is activated when a tilt of 1½° is reached. The height of the structure is 135 m, the circumference is 424 m and the weight of the wheel and capsules is 2100 tonnes. There are 80 cables with a total length of 6 km.

During operation the wheel turns continuously with a perimeter speed of 0.26 m/s, enabling easy access and exit from the capsule, and the ride, or 'flight' takes 30 minutes. Power is supplied by a LEB substation and standby power is supplied by a 1000 kVA generator.

British Airways London Eye

The Millennium Wheel, or *The London Eye*, was given planning permission by the London Borough of Lambeth to operate for five years, after which its future is uncertain.

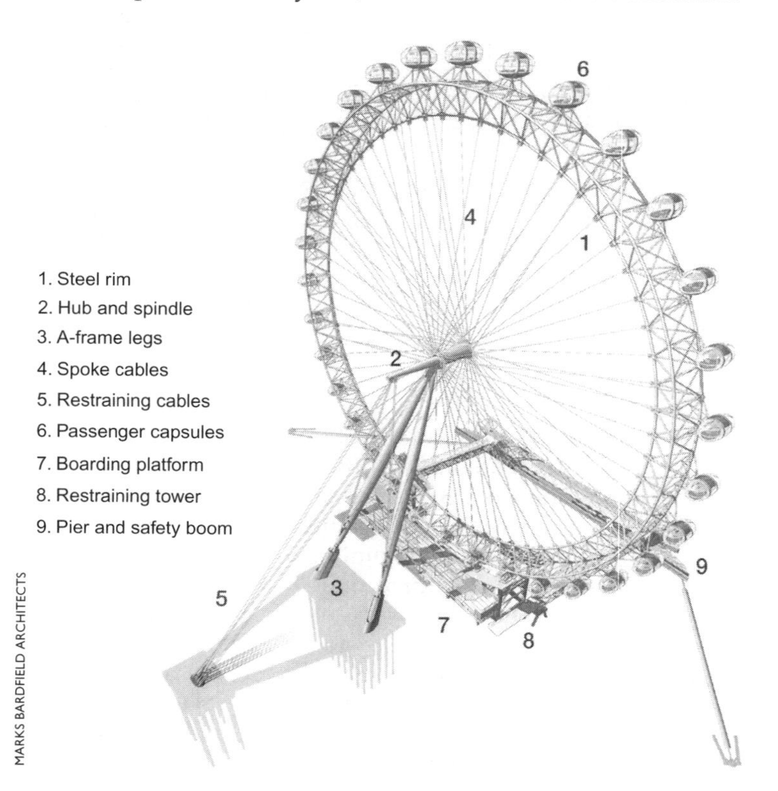

British Airways London Eye: lifting the wheel

Commerce

32. Millbank Tower

HEW 2362
TQ 302 787

When completed in the autumn of 1963 this structure was the highest reinforced-concrete office building in Britain. The development, on a 3½-acre site, comprises the 34-storey tower, an eight-storey Y-shaped office block and a 12-storey block of flats containing 80 dwellings. The development provides approximately 350 000 sq. ft of office space.

The tower rises 387 ft above ground level and the plan can be enclosed by a rectangle with sides approximately 120 ft by 110 ft, and it was built without the use of traditional independent scaffolding. The foundations comprise 163 piles capped by a raft. The structure was designed to withstand a sustained wind velocity of 75 miles per hour and the horizontal forces are resisted entirely by the reinforced-concrete core. The tight construction programme required the demolition of existing buildings on site to begin in January 1959, the excavation and foundations to bring the ground floor to completion in December 1959, the tower to be structurally complete by October 1961, and the boiler and main services to be in working order by November 1961. The external cladding is in stainless steel and glass.

An interesting and unusual feature of the project was the need to provide a Crown property known as the *Speaker's Stables*. Speakers of the House of Commons had traditionally enjoyed the privilege of stabling their horses and carriages, and the new 'stables' provide car parking and associated facilities.

The development was by Legal and General Assurance for the clients, Vickers and others. The Engineer was G. W. Kirkland of R. Travers Morgan & Partners, the architects were Ronald Ward & Partners and the main contractors were John Mowlem & Co. Ltd.

KIRKLAND, G. W. and FOUNTAIN, D. M. The Millbank Tower. *Consult. Engr*, March 1962, 290–93.

33. National Westminster Tower

HEW 2374
TQ 331 813

The 200 m high National Westminster Tower between *Old Broad Street* and *Bishopsgate* in the City, built 1971–81, was the tallest building in London until the completion of the Canary Wharf tower in 1991. The consulting engineers were Pell Frischmann, with Mowlem as the main contractor. Planning for the site began in the early 1960s, but it was not until 1968 that a development permit was issued, and the architects R. Seifert and Partners were appointed. The complexity of the site and the mechanical services requirements precluded a simple structural solution, and the design of the foundations for such a large structure in London Clay demanded careful consideration. The design was therefore accompanied by an extensive research programme, including structural modelling, an investigation of the dynamic stability and a finite element analysis of the foundation design. Compared with international practice some of the architectural features and construction methods were more appropriate to the previous decade. Although the low speed of erection of the floors lagged behind achievement elsewhere, it represented a major step forward in the British context. The extensive reinforcement in the core wall led to the development of a new method of slipforming in concrete—one example of the considerable challenge the design posed to the engineers and contractors involved. Construction was associated with a range of instrumentation with a view to benefiting the construction industry as a whole.

The tower is built on a massive foundation raft, 54 m in diameter and 4.5 m thick, which sits on 375 bored piles, each 25 m deep. On top of this raft a plinth supports the central core of the tower. The central core wall supports three cantilevered, reinforced-concrete, cellular beams which surround the core, giving the tower an irregular 12-sided shape in plan. Above these the core is surrounded by structural steel framework and lightweight composite floors.

The building suffered extensive damage in a terrorist attack in the City in 1993. Staff were relocated, and although the elevations have been preserved, restoration undertaken in the 1990s provided a new glazed fore building with a giant upswept canopy.

FRISCHMANN, W. W., SAINSBURY, R. N., *et al.* National Westminster Tower. *ICE Proc.*, 1983, **74**, 387–494.

34. Hop and Malt Exchange, Southwark Street

HEW 2321
TQ 326 802

Situated close to the Borough Market, the Exchange dealt with the hop, malt and seed trades and was well placed in relation to the Kent hop fields, with a rail connection to London Bridge Station. The architect was R. H. Moore and the first stone was laid on 31 August 1866. The Exchange floor in the centre of the building is 80 ft long by 50 ft wide. The original roof had iron lattice arch ribs of 25 ft radius with lattice purlins. The ribs were 3 ft deep at the springing and 18 in. deep at the crown, and were bolted to a cast-iron cornice. The merchants' offices (over 100 in total) were arranged on four floors, with decorative cast-iron balconies projecting into the four sides of the trading area. The arched roof was destroyed by fire in 1920 and a flat roof was substituted at cornice level, but the decorative balconies and wrought-iron entrance gates survive, together with the cast-iron columns on the *Southwark Street* facade. The building was opened in 1867 and is still in use as office accommodation.

The Builder, 5 October 1867, 730–33.

35. Smithfield Market

HEW 2363
TQ 318 817

In 1174 the market here was described as 'A smooth field where every Friday there is a celebrated rendezvous of fine horses to be sold', and the site became known as 'Smooth-field' for many years before becoming *Smithfield.* In 1638 the Corporation of London established a cattle market on the site and through the centuries this was much complained of by local inhabitants. The present central building was built in 1866–67 by Sir Horace Jones the City Architect. The building is 631 ft long by 246 ft wide with a central *Grand Avenue* 50 ft wide. The timber queen-post roof over the avenue has wrought-iron straps and cast-iron decorative roundels. Beneath the meat market building a 4-acre goods station, built in 1862–65, connected the market with the Metropolitan Railway. The engineer for the work was John Fowler of the Great Western Railway, who undertook the construction of the line jointly with the Metropolitan. The underground structure comprises twenty 240-ft span girders supported on 180 columns with brick jack-arches forming the roof. Hydraulic lifts with accumulators connected the station to the market above. The spiral ramp provided for horse-drawn traffic now gives access to an underground car park.

The original poultry market of 1873–75 designed by Jones was destroyed by fire in 1958. The roof of the new *Poultry Market* is formed of an elliptical paraboloid concrete shell of rectangular plan 222 ft by 127 ft 9 in. With such a large shell the designers felt that a model test was advisable. The rise of the shell is 30 ft from the corners to the centre. The roof is supported by columns along all four sides, providing clear space beneath. The shell roof is of reinforced concrete, generally 3 in. thick increasing to $6\frac{3}{4}$ in. at the edge zones. Prestressed concrete edge beams, using the Gifford–Burrow anchorage and $1\frac{1}{8}$ in. diameter strand cables, stiffen the edge of the shell. The roof is covered with copper sheeting laid on an insulation of lightweight screed. A number of holes 5 ft in diameter are fitted with glass domes to admit light. Construction began in January 1960 and the scaffold for the dome began to rise in April 1961. The prestressing of the roof was completed at the end of September 1962 and the official opening took place in May 1963.

The architects were Sir Thomas Bennet & Son, the Consulting Engineers were Ove Arup & Partners, and the main contractors were Sir Robert McAlpine & Sons Ltd.

36. Michelin Building

HEW 2322
TQ 273 787

This exotic Edwardian building faces the *Fulham Road* and is bounded by *Sloane Avenue* and *Lucan Place*. It was built for the Michelin Tyre Company to the architectural design of Francois Espinasse (1880–1925). The engineers were L. G. Mouchel & Partners. The first drawing by Espinasse was made in October 1909, the final designs in

Michelin Building

WENDIE TEPPETT

April 1910, and the building was completed in December 1910. The extravagant external decoration conceals a fine example of an early Hennebique ferro-concrete structure. The internal concrete structure comprises interior columns, lintels, floor and roof beams, and solid and hollow block floors and roofs. When originally completed the company had space at the rear of the building reserved for future expansion, and this was used in 1912. In 1922 further development took place, carried out by the Considere Company. In April 1969 the front (and oldest) part of the building was Listed Grade II. By 1985 the Michelin Tyre Company decided to relocate to Harrow-on-the-Hill, and in August of that year the building was bought by Sir Terence Conran and Paul Hamlyn who acquired planning permission for change of use in January 1986. The conversion of the building required 1300 separate detail drawings to enable 60 steel columns to be threaded through the existing floors to support a new fourth floor. The original features have been retained and restored. The building now comprises offices, retail shops and a restaurant.

HITCHMOUGH, W. *The Michelin Building.* Conran Octopus, London, 1987.

Government Buildings

37. Houses of Parliament

HEW 2323
TQ 303 795

The old Palace of Westminster was destroyed by fire on the night of 16 October 1834. The resulting architectural design competition was won by Charles Barry for his design of the present Houses of Parliament. The new building posed several engineering problems, notably the foundations, the need for a fire-resistant structure, the desire to reclaim land from the river for the terrace, and the provision of such building services as heating, ventilation and drainage. In fact, the present building can be regarded as the first where the services requirements materially impinged on the architectural design. Walker & Burges were appointed consulting engineers for the large cofferdam in which would be built the river wall and terrace. The contract for building the 920 ft long cofferdam was let to J. & H. Lee for £24 195. Building took

ILLUSTRATED LONDON NEWS, 1842, **7**, 104

Houses of Parliament under construction,

place between 1 September 1837 and 24 December 1838 and the cofferdam remained in position until 1849. The outer wall comprised a double row of 12 in. square piles, 36 ft long, with puddled clay between. The river wall was 7 ft 6 in. thick at the base and 5 ft at the top, built to a face radius of 100 ft. The terrace itself was formed of 26 000 cu. yd of mass concrete, and the building sits on a 7-acre mass concrete raft with an average thickness of 5 ft. The raft is thickened beneath the *Victoria Tower*, the *Central Spire* and the *Clock Tower*. These three vertical elements in the building were dictated by the consideration of the ventilation of the building. The general building contractors were Grissell and Peto, who made several innovations in their square-timber bolted scaffolding with travelling bridge cranes. The basement is formed of brick piers, based on the concrete raft and groined brick vaulting. The floor and raked seating of both houses is supported by columns and raking girders of cast iron. The pitched roof structure of both chambers was formed of cast-iron trusses clad with cast-iron plates, screwed to the rafters. The roof of the Commons was destroyed in the Second World War, but the Lords roof survives as originally constructed.

PORT, M. H. (ed.). *The Houses of Parliament*. Yale University Press, New Haven, CT, 1976, see chaps 10 and 11, 195–231.

DALRYMPLE, G. S. A description of the coffre dam at the ... Houses of Parliament. *Min. Proc. Instn Civ. Engrs*, 1840, **2**, 18–19.

38. Westminster Hall Roof

HEW 473
TQ 302 796

This historic roof, which survived the Houses of Parliament fire of 1834, is carried on a superb timber hammer-beam structure. The original hall was built under William II at the end of the eleventh century. Rebuilding began in 1394 during Richard II's reign but was not completed until 1402, three years after his deposition. It is a pre-eminent example of Gothic palatial architecture. The internal dimensions are 239 ft long by 68 ft wide, making the span of the hammer-beam roof one of the largest in Europe, and 50% wider than any previous hall roof in England. The roof carpentry is the work of Hugh Herland and is worthy of close inspection. The oak roof trusses are spaced at 20 ft centres and form 12 bays. During the eighteenth and nineteenth centuries all the external magnesian limestone and much of the internal stonework was renewed. In the early years of the twentieth century the roof was repaired and strengthened by steel reinforcement. Recent structural analysis and 1:10 scale model tests with resistance wire strain gauges have helped engineers to understand the manner in which the timber trusses distribute their load.

YEOMANS, D. T. (ed.). *The Development of Timber as a Structural Material.* Ashgate, Aldershot, 1999, 127–76.

ROYAL COMMISSION ON THE HISTORICAL MONUMENTS OF ENGLAND. *London: West London*, vol. II. HMSO, London, 1925.

39. Henry VIII's Wine Cellar, Whitehall

HEW 797
TQ 303 801

Within the Ministry of Defence buildings bounded by *Whitehall*, *Horse Guards Avenue*, *Richmond Terrace* and the *Victoria Embankment* there is a brick vaulted chamber which was part of the old Palace of Whitehall. Its survival is an interesting case study of the problems facing the civil engineer when moving an historic medieval structure to a new site. The structure in question had been used as a wine store by Henry VIII and is 62 ft long, 32 ft wide and 20 ft high and weighs about 1000 tons. The cut brick ribbed vaults are supported on stone corbels and by a central row of four octagonal stone columns.

Work on site for the new Ministry of Defence building was begun in 1939 and the Government had already agreed that the vaulted structure should be preserved. Just before the War the wine cellar was being strengthened with reinforced-concrete portal frames to withstand future underpinning in connection with the new Government building works. With the onset of war work ceased in 1940 and the wine cellar remained in place throughout the War.

In 1946 work on the office building restarted and the question of the wine vault was re-examined. It was sited on the ground floor of the new building and was an obstruction to the planned layout. It was decided to move it to the sub-basement level—making it a vault in the usual sense. This involved moving the whole brick box structure 43 ft 6 in. laterally to the west onto a prepared steel staging, lowering it 18 ft 9 in., and then rolling it back almost to its original position. The traversing was done by a drag plate and turnbuckles. The lowering was done by using 64 bottle screw-jacks, each with a full travel of 10 in. Sixteen men, deployed in pairs, turned each jack in sequence through one-eighth of a turn ($\frac{1}{16}$ in.). After the first full 10 in. of lowering, a second set of jacks was installed alongside the original jacks. The greatest rate of lowering achieved was 10 in. in 6 hours, and the whole movement took four months. The wine cellar was moved without any damage to the historic structure.

The new building is steel framed with solid reinforced-concrete floors and the external walls are of 13½ in. brickwork faced with Portland stone. The consulting engineers were R. Travers Morgan, the foundation contractors were Trollope and Colls, and Dorman Long & Co. Ltd. were the steelwork contractors.

WHITE, L. S. and GARDNER, G. A. Government Offices, Whitehall Gardens. The special problem of the re-siting of an historic building. *J. Instn Civ. Engrs*, 1949–50, **34**, 222–241.

PIKE, C. F. New Government offices, Whitehall Gardens. *Struct. Engr*, April 1948, 218–238.

40. Trafalgar Square Waterworks

HEW 2324
TQ 298 806

This well pumping station is situated in *Orange Street* and was built to supply water to Government offices in

Whitehall, the new Houses of Parliament and the fountains in Trafalgar Square. A contract for well-sinking, tunnelling and the supply of the pumping machinery was awarded, in January 1844, to Easton & Amos of Southwark, and the works were finished in December 1844. A well was sunk through the clay down into the chalk to a depth of 300 ft. A second well was sunk in front of the National Gallery. Two steam engines were installed, the principal one being of the high-pressure condensing Cornish type with a 30 in. diameter steam cylinder, lifting 100 gallons of water per minute to a tank in the top of the 54 ft tower. The second engine was on standby. By means of a tunnel under the National Gallery condenser circulating water supplied the fountains, was thereby cooled, and returned to the condenser. In December 1847 Buckingham Palace was also supplied from the works. The works were further extended in 1849 by sinking another well in the engine house and adding a 60 hp Woolf compound engine. In 1852 the tank tower was raised by 20 ft to supply tanks in the roof of the Houses of Parliament. The wells are now pumped by electricity and the buildings and tank tower remain.

AMOS, C. E. On the Government waterworks in Trafalgar Square. *Min. Proc. Instn Civ. Engrs*, 1859, **19**, 21.

41. Cabinet War Rooms, Whitehall

HEW 2325
TQ 299 797

Down Street underground station on the Piccadilly Line, between Hyde Park Corner and Green Park Stations, was closed in the early 1930s. With the imminent onset of the Second World War, Winston Churchill and his wartime Cabinet used Down Street Station while the Cabinet War Rooms at the end of *King Charles Street* were being prepared. The Cabinet War Rooms, situated in a former Government storage basement under the Treasury buildings, became the seat of operations in August 1939, just before the War began. Not surprisingly, it is somewhat difficult to obtain details of the work, but we do know that the engineer for the project was Brigadier James Orr, who received an OBE for his work. The work comprised a massive concrete slab, acting as a protective shield, and filling the ground floor of the building above.

The slab was supported on timber frames and riveted steel girders, which are clearly visible. The subterranean system covered about 6 acres and comprised several tunnels. The two principal tunnels run north and south. The southern tunnel connects the War Rooms with the Government Citadel in *Marsham Street*. The Citadel was built within the base of the Horseferry gasholders of the Gas, Light & Coke Company. The rotunda building can still be seen from *Monck Street*, Westminster. The northern tunnel links the War Rooms with the Government telephone exchange at Craig's Court near Charing Cross Station. The Cabinet War Rooms are open to the public as part of the Imperial War Museum.

42. Institution of Civil Engineers, Great George Street, Westminster

HEW 2327
TQ 299 796

Several interesting buildings related to the education and training of civil engineers survive in London and the Thames Valley, although others have been demolished. An unusual, but important, attempt to educate civil engineers in a private establishment was the founding of the Putney College for Civil Engineers. It opened originally in May 1840 at Gordon House, Kentish Town, but re-opened in August of that year at Putney. The 1851 Census reveals that it had 35 students, a head, three academic staff and 13 domestic staff. The civil engineering profession was sceptical about the college and preferred the system of pupilage. The *Civil Engineer and Architect's Journal* of September 1840 went so far as to say 'the very fact of having attended this college would go far to exclude those who had done so from the profession'. It was not a financial success and closed in 1857 and the buildings were demolished.

The need for consulting civil engineers to be located near the Houses of Parliament led to a great aggregation of engineering practices in Westminster, notably in *Great George Street*, *Duke Street* and, later, *Victoria Street*. An engineer working on surveys, reference books, plans and estimates for a client seeking a Bill in Parliament would have to meet the dreaded annual deadline of midnight on 30 November, when the shutters at the Bill

The Institution of
Civil Engineers

DENIS SMITH

Office would close. If this deadline were missed, a whole Parliamentary Session would be lost for the client's project. It is not surprising, therefore, that when a civil engineering institution was proposed in 1818, premises would be sought in the locality. The fledgling Institution of Civil Engineers, the world's first professional engineering institution, rented its first headquarters accommodation, from 1820 to 1832, at 15 *Buckingham Street*, Adelphi. This comprised one room in the house of the Rev. James Harris. From 1832 to 1839 they were housed in No. 1 *Cannon Row*, on a seven-year lease at £80 per year. The Institution acquired a house at 25 *Great George Street* in December 1837 and sublet No. 1 *Cannon Row* until the end of the lease.

25 Great George Street was to be the headquarters building of the Institution until 1910. Improvements and enlargements were carried out to the premises in 1846, 1868 and 1872 to the designs of Matthew Digby Wyatt. In 1894–95 a new building was erected on the sites of 24–26 Great George Street, all owned by the Institution, designed by the architect Charles Barry (junior). This purpose-built building, decorated to celebrate the development of civil engineering, contained many internal features which were subsequently transferred to the present building erected in 1910–13 on the site of 1–7 Great George Street. This move was a consequence of the Government Offices Act of 1908.

The present building is a steel-framed structure, one of the earliest to be erected in London following the changes to the London building by-laws in 1909. A competition, won by James Miller, was held for the design of the building, which reflects the prestige of the Institution and the engineering profession on the eve of the First World War. Ferdinand Huddleston acted as structural engineer for the steelwork, and the contractors were Mowlem. Further work was done in the 1930s when the lease of 1 Great George Street was finally purchased and the north-west corner completed. A major refurbishment, including an extension on the Prince's Mews elevation, was completed in 1990 to the design of the Building Design Partnership, with the main contractors being Holloway, White, Allom.

The building today houses extensive conference facilities, Institution offices and a Library which, with its archive collection, is one of the world's leading sources of information on the development of civil engineering down to the present day. A collection of works of art, including the world's largest collection of engineering portraits, is displayed in the building, the main atria of which are decorated with the names of the great British engineers of the past 400 years.

43. The Royal Indian Engineering College, Cooper's Hill

HEW 2328
TQ 995 721

The East India Company trained its own engineers at a college in Addiscombe, south London, which was opened in 1809 and closed after the Indian Mutiny of 1857. The Government eventually decided to replace it with a civil engineering college at a site on Cooper's Hill near Runnymede.

The estate was bought in 1870 for use as the Royal Indian Engineering College and the property was converted under the direction of Sir Matthew Digby Wyatt. There were four classrooms, a library, a lecture theatre, a model room, a dining hall and individual bedrooms for each student—100 at first, rising to 150. The President was housed in what is now called *President Hall,* and corridors retain names with Indian associations such as *Clive Corridor* and *Warren Hastings Corridor.* The formal opening of the Royal Indian Engineering College was performed on 5 August 1872 by the Duke of Argyle. The students entered the service of the Government of India's Public Works Department, mostly as engineers, at a starting salary of £420. The college has an important place in the development of engineering education and attracted an eminent staff. The first engineering appointment was that of Callcott Reilly as Professor of Construction. W. C. Unwin was appointed Professor of Hydraulics and Mechanical Engineering in 1872 and left in 1885 to join the City and Guilds College in South Kensington. The college closed in 1906 when the work was transferred to India in a college using the name *Cooper's Hill.*

The original buildings remained empty for some years, until January 1911 when it became the family home of Baroness Cheylesmore until 1925. By 1938, with the threat of war, the London County Council bought the estate as an emergency headquarters and they moved there in 1939. In 1946 it became the Cooper's Hill Emergency Teacher's Training College, which closed in 1951. The Teacher Training Department of Shoreditch Technical College transferred to Cooper's Hill in 1951. Brunel University became tenants of the estate in August 1980.

44. Crystal Palace School of Practical Engineering, Sydenham

HEW 2329
TQ 338 707

Sited at the top of *Anerley Hill* on the Sydenham site of the Crystal Palace, this school was founded in 1872 by the Crystal Palace Company and was established to provide students of any branch of engineering with a thorough practical and theoretical instruction in the rudiments of the profession. Pupils were usually

Crystal Palace School of Practical Engineering, Sydenham

DENIS SMITH

DENIS SMITH

Royal Military Academy, Woolwich

admitted at the age of 16–17 years and studied mechanical engineering in the first year and civil engineering in the second. An advertisement from 1919 states that 'The school is approved by the Council of the Institution of Civil Engineers'.

The surviving building is of London stock brick with a hipped-gable slate roof. It is adjacent to the base of I. K. Brunel's south water tower, which survived the fire of 1936 but was demolished during the Second World War for security reasons. The school building now houses the Museum of the Crystal Palace Foundation and is the last Victorian building surviving on the site.

45. Royal Military Academy, Woolwich

HEW 2330
TQ 431 772

Sited between *Academy Road* and *Red Lion Lane*, Woolwich, the Royal Military Academy was first established within the Royal Arsenal in 1719. It was moved to its present site on Woolwich Common in 1805 when James Wyatt designed the impressive building. Its Gothic-style north front is 720 ft long with a central block and two side blocks.

The early members of staff included such eminent men as Charles Hutton, Olinthus Gregory and Professor Peter Barlow, a scientist and mathematician with a European-wide reputation who was much involved with civil engineering questions. Peter Barlow carried out a series of materials tests on the testing machine in Woolwich Dockyard, which led to the publication, in 1817, of his book *An Essay on the Strength and Stress of Timber*, which became a standard engineering textbook on materials for most of the nineteenth century. Many important military engineers who became involved in civil engineering matters were educated here. The Academy buildings are Listed Grade II*.

46. University College, London

HEW 2338
TQ 296 823

The University of London was founded in 1826. The College in *Gower Street* became known as *University College London* on its federation with King's College in 1836. University College was one of the pioneering educational establishments to teach civil engineering. Before engineering, as such, was taught, Dr. Dionysus Lardner was appointed in 1827 as the first Professor of Natural Philosophy. The first Professor of Civil Engineering was Charles Blacker Vignoles, who was appointed in 1841. But the man who made a distinctive contribution to engineering education was Alexander Kennedy, who was appointed Professor in 1874 and was the first to introduce an engineering laboratory in a British university. A college policy document of 1878 states that 'The formation is contemplated in a suitable quarter of the premises of a new Engineering Laboratory. The Council and Senate are desirous of making University College one of the chief places of Scientific Education for the Engineer'. The laboratory was located in a surviving basement area of the college and included a 50-ton materials testing machine designed by Kennedy. This testing machine was in use until 1958.

King's College was the second college of the university of London and was founded in 1828. A Department of Engineering was formed in 1831.

BISHOP, R. E. D. Alexander Kennedy: the elegant innovator. *Trans. Newcomen Soc.*, 1974–75/1975–76, **47**, 1–8.

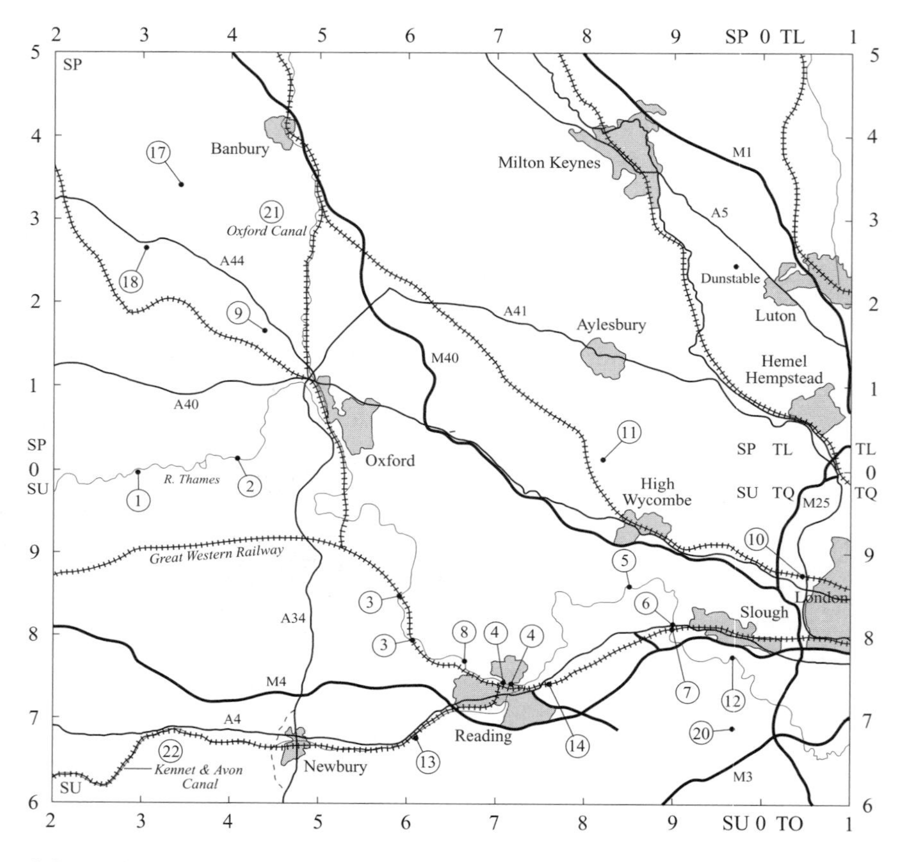

1. Radcot Bridge
2. Newbridge
3. Basildon and Moulsford Railway Bridges
4. Reading and Caversham Bridges
5. Marlow Bridge
6. Maidenhead Road Bridge
7. Maidenhead Railway Bridge
8. Mapledurham Watermill
9. Grand Bridge, Blenheim Palace, Woodstock
10. Denham Court Lenticular Bridge
11. Lacey Green Windmill
12. Windsor Town Bridge
13. Great Western Railway
14. Sonning Cutting
15. River Thames Bridge, Great Western Railway, Windsor Branch
16. Bourne End Viaduct
17. Hook Norton Brewery
18. Bliss Tweed Mill
19. River Thames Navigation
20. Virginia Water
21. Oxford Canal (Banbury to Oxford)
22. Kennet Navigation and Kennet and Avon Canal

9. The Thames Valley

There are two main topographical features of the area, the limestone of the Cotswolds to the north and the chalk hills of the Berkshire Downs and Chilterns which cut across the centre. Sandwiched between are the clays of the Oxford Plain and to the south of the Chilterns lie the Reading and London Clays. The limestone has proved excellent for building, giving the characteristic warm stone of so many Oxfordshire towns and villages. Elsewhere the clays provided the raw material for brick-making. There are few minerals. Haematite iron ore was found near Banbury, but was largely exhausted by the early part of the twentieth century. Until recently cement was made on the scarp of the Chilterns at Chinnor and Pitstone.

The River Thames has long been a main commercial artery, and dictated the development of part of the canal system—the Oxford Canal taking the course of the River Cherwell into Oxford. The Kennet and Avon Canal followed, or used, the Kennet in order to reach the navigable Thames at Reading. As the river also formed a considerable barrier, its bridges constituted vital links, some fought over and damaged during the Civil War. In view of their importance a number of these bridges are included in this chapter. Only those bridges on the upper reaches of the river, where stone was easily accessible, have lasted from medieval times. Elsewhere, although timber was freely available, wooden bridges were comparatively short-lived and longevity had to wait for a wider use of stone, brick, cast and wrought iron, and reinforced concrete.

A number of ancient tracks and Roman roads cross the area, but perhaps the best known is the *Great West Road* from London to Bath. It follows the Kennet valley through Newbury from Reading, whereas when planning a new mode of transport Brunel chose to take his Great Western Railway to the north of the Berkshire Downs by taking the line through the Goring gap. It was only seven years later, however, that he took the Berks & Hants Railway through the Kennet valley. From Didcot the railway was extended to Oxford in 1844 and to Banbury in 1850, and the line to Evesham was opened in 1853.

1. Radcot Bridge

HEW 1913
SU 285 993

Radcot Bridge carries the Witney to Faringdon Road and was built about the fourteenth century, on or near the site of an earlier bridge mentioned in a grant of land made by King Eadwig in 958. It is probably the oldest medieval bridge over the Thames.

The 12 ft wide bridge, which is all in stonework, has three arches, a centre span of 12 ft and side spans each of 10 ft. The side spans are sharply pointed and each has four ribs. The centre span has a flat, four-centred shape without ribs and was probably reconstructed at some time in its history, possibly connected with military activity. Scarcity of bridge crossings over the river gave the bridge a high strategic value, and Radcot has featured twice in history as a site for skirmishes. In 1387 the Earl of Oxford, in attempting to march on London, with troops raised in the north-west, was intercepted and beaten by the rebel Barons. During the Civil War, only a month before the battle of Naseby, Parliamentary forces were repelled by Lord Goring on his way to join the King at Oxford.

The narrow centre arch imposed a severe restriction on barge traffic in the eighteenth century, and in 1787 William Jessop advised that a new bypass cut be made on the north side, 220 yd long and complete with a new single-span bridge, leaving the old river bed to silt up. The original bridge is now a scheduled Ancient Monument and is Listed Grade I.

TOYNBEE, M. R. Radcot Bridge and Newbridge. *Oxoniensia*, 1949, **14**, 46–49.

JERVOISE, E. *The Ancient Bridges of the South of England*. Architectural Press, London, 1930, 2–3.

2. Newbridge

HEW 1914
SP 404 014

Newbridge lies on the road between Kingston and Witney and spans the River Thames at its confluence with the River Windrush. It was probably built in the fifteenth century and remains essentially as originally constructed. John Leland's description in the middle of the sixteenth century is clearly of the same bridge that can be seen today. It has six pointed arches with spans ranging

J. B. POWELL

Newbridge, River Thames

from 12 ft to 19 ft all in stonework. The four inner arches appear to have each had four reinforcing ribs, some of which have been cut away, presumably to ease navigation. Only the second arch from the left bank retains all four ribs. As in many early bridges, cutwaters were only provided on the upstream side, and in this case were carried up to parapet level to provide refuges for pedestrians. The width of the bridge is 15 ft between the parapets.

During the Civil War in May 1644, Newbridge was defended for the King from attacks by Waller, and although the first was beaten back, a few days later the defenders were overcome after Waller had used the stratagem of crossing by boats and making a surprise attack. In October of that year the bridge was damaged by Parliamentary forces during the Royalist retreat at Wallingford. The bridge is a scheduled Ancient Monument and is Listed Grade I.

DREDGE, J. *Thames Bridges from the Tower to the Source.* Engineering, London, 1897, 206–07.

3. Basildon and Moulsford Railway Bridges

Basildon:
HEW 1297
SU 606 796

Moulsford:
HEW 1298
SU 595 847

West of Reading Brunel kept the Great Western Railway to the south of the River Thames, squeezing the line between the river and the high ground at Pangbourne until both river and railway turned north. With the river coursing between the Chiltern Hills and the Berkshire Downs two crossings were required, the first at Basildon, 3 miles from Pangbourne, and the second beyond the Goring Gap at Moulsford.

In their original form, accommodating two broad-gauge tracks, both bridges were built 30 ft wide with four semi-elliptical arches in red brickwork and with Bath stone voussoir facings. The arches at Basildon were slightly skewed, with spans of 64 ft 4 in., each with a rise of 19 ft. Owing to the greater skew the Moulsford Bridge spans were 87 ft, the rise being 22 ft 7 in.

The quadrupling of the line from Taplow to Didcot was incorporated with the final broad gauge conversion to standard gauge in 1892. Both bridges were therefore required to be widened. The arch profiles set by Brunel were followed, but without the masonry embellishments of the originals. The Basildon Bridge widening of 26 ft

Basildon Rail Bridge

J. B. POWELL

abutted the original on the downstream side, whereas in the case of Moulsford Bridge the widening was on the upstream side, the halves of the bridge being separated by a gap of 9 ft 6 in. The engineer for the original bridges was I. K. Brunel and the contractor was W. Chadwick (for both original bridges).

MacDermot, E. T. and Clinker, C. R. *History of the Great Western Railway*. Ian Allan, London, 1964, vol. 1, 54; vol. 2, 205.

4. Reading and Caversham Bridges

Reading:
HEW 2179
SU 718 741

Caversham:
HEW 2180
SU 711 746

These two bridges over the Thames represent fine examples of reinforced concrete applying the Mouchel–Hennebique system of the 1920s. Of the pair, Reading Bridge was the first to be built, anticipating the opening of the replacement Caversham Bridge by nearly three years.

The construction in the one case and the reconstruction in the other arose from the Reading (Extension) Order of 1911, which conveyed most of Caversham to Reading Borough. Caversham Bridge was to be widened, and an additional bridge at least 10 ft wide was to be built over the river. With commendable good sense the Corporation decided that, if properly sited, a new vehicular bridge, 40 ft wide, would be of much greater

Reading Bridge

J. B. POWELL

benefit, not only to Caversham, but also giving better access to the Henley Road. In the case of Caversham the existing bridge dating from 1868 consisted of wrought-iron lattice trusses on cylindrical cast-iron piers. In addition to being a rather ugly bridge, its construction did not lend itself to widening, and it was therefore decided to reconstruct it. Additional parliamentary powers were sought to include all the changes arising from the decisions taken. These included a length of main road to the north of the proposed new bridge and property acquisitions on the north side of Caversham Bridge to enable the widening to be carried out. The Reading Corporation Act was passed in August 1913 and L. G. Mouchel and Partners were commissioned to design both bridges.

The requirement for Reading Bridge was for a single river span of 180 ft with side spans. Stepped concrete blocks 50 ft long and 45 ft wide were provided on each side of the river span to take the thrust of the four main arch ribs, each 4 ft 6 in. wide and 8 ft 3 in. apart. These ribs in turn supported the deck slab by means of spandrel columns of diminishing height towards the bridge crown. The Great War prevented construction at the time, but in April 1922 contractors Holloway Brothers started the works. The total cost, including river and approach works, amounted to nearly £70 000. Claimed at the time to be the largest bridge of its type in the country, it was opened on 3 October 1923.

Caversham has a long association with St. Anne, and a chapel in her memory stood on the island separating the two parts of the bridge. During the reconstruction in 1924, parts of the foundations of the chapel were uncovered and were removed by the Corporation for safekeeping. In the sixteenth century Leland described the structure as a 'great mayne bridge of tymbre' and noted that 'it restid most apon fundation of tymbre, and yn sum places of stone'. It was the decayed state of the successor of this bridge which prompted the reconstruction of 1868.

With Reading Bridge now open, Caversham could be closed to vehicles and a temporary bridge erected alongside to take services and pedestrians. Although more decorative, the design was similar to that of the earlier bridge. The south (Reading) span is 126 ft 4 in. and the north (Caversham) span 106 ft 4 in., the width between parapets being 56 ft. Again, the contractors were Holloway Brothers, who started work in March 1924 and completed in April 1926. The total cost was £71 000. At the official opening on 25 June 1926 plaques giving a brief history and details of construction were unveiled by HRH the Prince of Wales.

PHILLIPS, G. *Thames Crossings, Bridges Tunnels and Ferries*. David & Charles, Newton Abbot, 1981, 82–92.

DREDGE, J. *Thames Bridges from the Tower to the Source*. Engineering, London, 1897, 144–50.

5. Marlow Bridge

HEW 661
SU 851 861

In 1309, letters patent were granted to Gilbert de Clare, Earl of Gloucester and Hereford, granting pontage for four years to repair 'your bridge which is decayed and broken'. There were further grants up to 1399, the last to the town bailiffs. At this time the bridge was located a little to the east of the present bridge, opposite to which is now St. Peter Street. Except for Leland's confirmation that construction was in wood, little is known of the bridge until 1642, when the Parliamentary forces partly destroyed it. The cost of repairs was met by the issue of a

Marlow Bridge

J. B. POWELL

county rate. However, by 1789 the bridge had become unsafe and a public subscription, initiated by the Marquess of Buckingham, raised the sum of £18 000 for a new bridge, which was also built in wood on the same site. By 1828, however, this bridge also was in a state of complete disrepair. Maintenance revenues from lands in Marlow were woefully insufficient, and in order to finance a new bridge the Marlow Bridge Act (10 Geo.4) was passed on 14 May 1829. The Act gave Justices powers to raise funds by rates or private loans in addition to the sale of the revenue lands. Financial liability was apportioned four-fifths to Buckinghamshire and one-fifth to Berkshire, a ratio which arose from the position of the boundary within the river at the site of the first two bridges.

The newly constituted Bridge Committee rejected the conventional construction options of wood, stone and cast iron, and chose a suspension bridge. Although this was likely to be more expensive, the Committee may have thought it would give an imposing entry to the town and they could be seen to be in the forefront of technical advancement. The contractors were T. Clifford and Thomas Corby of Marlow, William Bond also of Marlow, and William Hazeldine of Shrewsbury. The designer, John Millington of Hammersmith, commenced the work, but on his resignation William Tierney Clark was appointed. He made alterations to the chains and piers, and when completed in 1832 the bridge looked very much as it does today. Final costs were £22 000, including the approach works.

In 1860 Easton, Amos and Sons of The Grove, Southwark, prepared a scheme for the replacement of the oak cross-beams by wrought-iron girders. In 1908 Marlow Urban District Council requested that the bridge be strengthened to carry 12 tons and, although some redecking was carried out in 1910, by 1924 a gross weight restriction of 5 tons had to be imposed. This move prompted discussions between the two counties to replace the bridge by a single arch in reinforced concrete after the manner of the recently completed Reading Bridge. These proposals came to nought, but when, in 1958, the weight limit was reduced to 2 tons it was decided to reconstruct the bridge back to a 5-ton limit. Local opinion was heavily in favour of retaining the original form of the bridge. The consultants Rendel, Palmer and Tritton prepared the scheme with steel chains, the main contract being awarded to Horseley Bridge and Thomas Piggot with Aubrey Watson as subcontractors. The total cost was £219 000.

The main span of the bridge is 227 ft 3 in. between pier centres, with side spans of 71 ft and 70 ft 4 in. on the Buckinghamshire and Berkshire sides, respectively. The carriageway is 18 ft wide and there are footways each side, both 4 ft wide.

SMITH, D. The works of William Tierney Clark 1783–1852: Civil Engineer of Hammersmith. *Trans. Newcomen Soc.*, 1991–92, **63**, 189–91.

WADSWORTH, H. J. and WATERHOUSE, A. Modern techniques and problems in the restoration of Marlow Suspension Bridge. *Proc. Instn Civ. Engrs*, 1967, **37**, 297–316.

DREWRY, C. S. *A Memoir on Suspension Bridges*. Institution of Civil Engineers, London, 1832.

6. Maidenhead Road Bridge

HEW 1881
SU 901 814

The road bridge across the Thames at Maidenhead was designed by Robert Taylor, architect of the King's Works, and built by John Townsend of Oxford. Work was started in 1772 and completed five years later. It replaced a narrow and much repaired wooden structure of late

Maidenhead Road Bridge

J. B. POWELL

thirteenth century origin sited close by. The 30 ft wide bridge has 13 semi-circular arches, seven in the main structure spanning the river and three in each of the approaches. The overall length of the bridge is 474 ft. The seven main arches are of Portland stone and vary in span from 35 ft at the centre to 28 ft 5 in. at the sides. The approach spans vary from 20 ft to 16 ft and are of brick construction except for the stone voussoirs and spandrel walls. The Act empowering construction of the new bridge provided for the alteration of the Buckinghamshire and Berkshire boundary line to ensure that the entire structure lay within the latter county. The bridge is Listed Grade I.

PHILLIPS, G. *Thames Crossings: Bridges, Tunnels and Ferries*. David & Charles, Newton Abbot, 1981, 110–15.

7. Maidenhead Railway Bridge

HEW 30
SU 902 810

While the first major structure on Brunel's railway was the Wharncliffe Viaduct crossing the Brent valley at Hanwell, a more formidable challenge presented itself at Maidenhead in crossing the Thames. In order to maintain the flat gradients Brunel had set himself, the bridge had to be aligned as low as possible while maintaining the navigational clearances required by the river conservators. He produced a masterpiece of two semi-elliptical spans in brickwork, each 128 ft long with a rise of only 24 ft 3 in., the centre pier being founded on a small midstream island. At the time they were built, between 1837 and 1838, they were the flattest bridge arches ever built and, as such, attracted much criticism and controversy concerning their stability. Indeed, the three lowest rings on the eastern span did separate for a distance of between 25 ft and 30 ft, probably owing to the mortar not having fully hardened. Remedial works were carried out and, after easing, Brunel instructed that the centres should be left in place over the winter, and there they remained until blown down in a storm. The contractor was Chadwick.

To accommodate the quadrupling of the line the bridge was widened on both sides by Sir John Fowler between 1890 and 1893, giving an overall width between the parapets of 57 ft 3 in. It was found that, owing to movements in the structure, the bridge soffits had

J. B. POWELL

Maidenhead Railway Bridge

become slightly distorted, but the Chief Engineer wished the new work to be built to true ellipses, and this requirement accounts for the discontinuity that can be clearly seen from beneath the bridge. In order to avoid any differential settlement relative to the old work, the foundation extensions were close piled and covered with a wooden grillage before being filled with concrete. There are four flood arches on each side, one of 21 ft and three of 28 ft span.

DREDGE, J. *Thames Bridges from the Tower to the Source*. Engineering, London, 1897, 123–25.

PUGSLEY, SIR A. (ed.). *The Works of Isambard Brunel*. Chapter. V: OWEN, J. J. B. *Arch Bridges*. Institution of Civil Engineers/University of Bristol, London/Bristol, 1976, 89–106.

8. Mapledurham Watermill

HEW 2150
SU 669 767

Mapledurham watermill is situated about 4½ miles upstream from Reading, and is the last working corn and grist mill on the River Thames. A 7 ft head of water drives a 12 ft diameter undershot wheel mounted on a 16 ft long horizontal oak shaft. The drive is transferred to a vertical shaft by a conventional pit wheel and wallower.

J. B. POWELL

Mapledurham Watermill

Above the wallower the wooden compass arm spur wheel is mounted on the shaft immediately beneath the stone floor and drives the two stone nuts, which in turn drive the two runner stones. Also on the stone floor a crown wheel drives the ancillary equipment by means of a shaft, pulleys and belting. There has been a mill here for 500 years. The buildings were extended during the mill's heyday in the seventeenth and eighteenth centuries and latterly to satisfy the then current London fashion for white bread. The flour was taken down river to London by barge. Today visitors can still buy flour ground at the mill on open days.

9. Grand Bridge, Blenheim Palace, Woodstock, and New Bridge, Bladon

Grand Bridge: HEW 2058 SP 439 164

New Bridge: HEW 2057 SP 443 152

The Grand Bridge was designed by John Vanbrugh in 1706 as an integral part of his plan for Blenheim Palace, which was to be the gift of a grateful nation to John Churchill, First Duke of Marlborough, victor of the battle of Blenheim. Construction of the bridge by the contractors, Bartholemew Peisley Sr. and John Townsend, commenced in 1708, and the main arch was turned in 1710.

J. B. POWELL

Grand Bridge, Blenheim Palace

However, following the Duke's fall from royal favour, all Treasury payments ceased and the work stopped, leaving the work unfinished. In 1721, five years after Vanbrugh's resignation from the project, work on the bridge was resumed at Marlborough's expense by John Townsend's son William and Bartholemew Peisley Jr., who completed the parapets and wing walls in their present form and canalised the River Glyme through the main arch.

In 1764 Capability Brown was engaged to landscape the Blenheim Estate and among other works proceeded to dam the Glyme to form a lake, raising the water level by some 13 ft, flooding the apartments embodied in the bridge and partially submerging the three arches.

New Bridge was constructed in masonry with three segmental arches, a strong string course and balustraded parapets in the manner characteristic of its designer, William Chambers. It carries an estate road to Bladon over the River Glyme within the grounds of the Palace and was built by Cheer in 1773. The main span is 34 ft and each side span 26 ft 6 in., the width between parapets being 17 ft.

DOWNES, K. *Vanbrugh*. Zwemmer, London, 1977, 55–123.

10. Denham Court Lenticular Bridge

HEW 1919
TQ 052 873

This graceful structure was built within the Denham Court Estate, probably in about 1870, as one of three footbridge crossings of the River Colne, the other two being suspension bridges. It is a fine example of this elegant form of construction now rare in England but more common in Scotland. Built in wrought iron, with a span of 61 ft, it consists of upper compression T-members supported from round bars in catenary as lower tension members. The supports consist of seven V-shaped spandrels increasing in size to mid-span to form the characteristic lentil-shaped truss. Each spandrel has a circular stiffener surmounted by an inverted member of segmental shape. At their ends the main members are anchored together within a casting that is bolted down to the abutments. The width between parapets is 4 ft 6 in.

The delicate parapets consist of groups of three posts of flat section with four horizontal rails, the lower three being round bars and the upper one, of half-round section, forming the handrail.

Following the failure of one of the wrought-iron tension bars the bridge was completely refurbished in 1991–92 using as much of the original material as

Denham Court Lenticular Bridge

J. B. POWELL

possible. However, steel bar was substituted for wrought iron in the tension members to increase strength. The structure is Listed Grade II*.

11. Lacey Green Windmill

HEW 2158
SP 819 007

The mill was originally built in Chesham, although the site is unknown. It is the oldest surviving smock mill in the country and its date is commonly given as 1650, although the precise provenance is difficult to substantiate. Certainly the main working parts are of great age

J. B. POWELL

Lacey Green Windmill

and have been described as being of a type which would have been approaching obsolescence in the seventeenth century. In particular, the brake wheel with its six radial spokes is all in wood, including the tyre and cogs. The circular wind shaft is of 25 in. diameter at the brake wheel and tapers severely to the end bearing. The wallower, located on the third floor, is also in wood and is wedged to the large vertical shaft. The machinery on the second floor is driven from an iron crown wheel and pinion to belted drives. On the first floor the great spur wheel with its compass arm spokes engages with iron-bound, solid wood, stone nuts driving two pairs of stones. Externally, luffing was carried out by a pulley wheel and chain attached to the curb gearing. In its working state the mill had one pair of common and one pair of patent anticlockwise sails. In 1821, by order of the Duke of Buckingham, the mill was moved to its present position facing west on a plateau of the steep scarp of the Chilterns. The ground floor is partly below ground and the octagonal superstructure is mounted on a dwarf brick wall. The mill last worked in about 1915. By 1930 it was very dilapidated but it was strengthened and renovated in 1935–36. However, in 1964 it was reported that it was not repairable and a suggestion was made that it should be rebuilt as a tower mill, and by so doing safeguard the machinery. Fortunately, in 1971, and by arrangement with the owner, the Chiltern Society carried out extensive refurbishment lasting several years, including eradication of a major twist in the framework which had followed degeneration of the lower ends of some of the main timbers. The sails and machinery are worked occasionally, but not the stones.

SCIENCE MUSEUM LIBRARY. *Windmills of Buckinghamshire*. Science Museum, London, 1977, vol. 1/5, 106–10.

BROWN, R. J. *Windmills of England*. Robert Hall, London, 1976, 46–47.

12. Windsor Town Bridge

HEW 913
SU 967 772

The foundation stone of the present Town Bridge across the Thames at Windsor was laid by the Duke of York on 10 July 1822 and the bridge opened almost two years later on 1 June 1824. It replaced an earlier and much repaired structure, recorded as far back as 1172, although

J. B. POWELL

Windsor Town Bridge

probably built even earlier. The bridge, on granite piers, has a central span of 58 ft and two side spans of 44 ft. Each span has seven cast-iron ribs, with a rise in the centre span of 6 ft 6 in. and in the side spans of 5 ft 6 in. The width of the bridge between parapets is 25 ft 6 in. The bridge was designed by Charles Hollis, who at the time was working as Engineer under Jeffry Wyatt on the restoration of Windsor Castle. The contract was awarded to William Moore, but on his death in 1823 the work was completed by his Executor, Baldock—one of the ironmasters supplying the arches for the bridge. Tolls continued to be paid, much to the annoyance of the users, and it was only on 1 December 1898, over 70 years later, that the bridge was finally freed from tolls after a three-year legal wrangle through the courts.

Following the opening of a bypass in 1966, which included the construction of Elizabeth Bridge over the Thames, the Town Bridge was closed to vehicular traffic in April 1970, and reserved for pedestrian use only. The bridge is Listed Grade II.

READING RECORD OFFICE. *Contemporary Bridge Committee Minutes.*

PHILLIPS, G. *Thames Crossings: Bridges, Tunnels and Ferries.* David & Charles, Newton Abbot, 1981, 118–23.

13. Great Western Railway

HEW 2268
Paddington to Swindon:
TQ 265 815 to SU 149 856

Following the construction of the Liverpool & Manchester Railway it became clear to the merchants of Bristol that their declining shipping trade would be at an even greater disadvantage compared with Liverpool unless they too had similar access to their markets, notably London. After earlier half-hearted attempts had come to nothing, at the end of 1832 a consortium of commercial and corporate interests in Bristol agreed to promote a railway between Bristol and London. I. K. Brunel, then 27 years old, was appointed Engineer, reporting to two Boards of Directors, in Bristol and London.

With the site of a permanent terminus in London left open, Brunel planned his route with minimum gradients westwards from London through West Drayton and Slough, crossing the Thames at Maidenhead before turning towards Reading. He could then choose to swing the line north of the Berkshire Downs, through the Vale of the White Horse and to Swindon, or through the Kennet Valley with its much higher summit. He chose the former and, while there were no major towns between Reading and Bath, the route offered future access to the Midlands and South Wales.

The line to Swindon was opened in six stages, the first being from London to Taplow in June 1838, and the last to Swindon in December 1840. There were branch lines to Oxford (1844), Windsor (1849) and Henley-on-Thames (1857). Two, much larger, subsidiary companies were the Oxford & Rugby Railway to Banbury (1850) and the Berks & Hants Railway, which took the line through the Kennet Valley to Newbury and Hungerford (1847) and included the line to Basingstoke (1848). A number of independent lines were built with connections to the main line, most of them, for reasons of financial stringency, being operated by the Great Western Railway until conditions were ripe for full absorption. The Great Western Railway gave financial assistance to the Oxford, Worcester & Wolverhampton Railway, but it was always a refractory child of the parent company.

Apart from the building of Paddington Station (see p. 169), the major works on this part of the line were the crossing of the River Brent at Hanwell by the eight-span

Wharncliffe Viaduct, Sonning Cutting and three crossings of the River Thames. The most notable of these was at Maidenhead, the other two being upstream of Reading at Basildon and Cholsey (Moulsford).

The track gauge of 7 ft (later widened to 7 ft ¼ in.) distinguished Brunel's railway from other systems. He convinced his Boards that the extra width would offer greater speeds and safety, together with smoother travel. His views were at the same time both visionary and short-sighted. In his vision he saw his transport system extending far beyond Britain, but he could not see that the owners of railways already constructed, and with a much greater mileage than his, would not countenance changing their own gauge. The various railway companies expanded and the broad and narrower gauge first met at Gloucester in 1845. It was now absolutely clear that there could be no exchange of vehicles or trains and that transhipment of goods and passengers would place great burdens of time, cost and inconvenience on the growing system. The Parliamentary Gauge Commission of 1846 recommended the adoption of a standard gauge and, although more broad-gauge lines were approved, the Great Western Railway became increasingly isolated. On some broad-gauge lines a third rail was added for mixed traffic, others were converted to standard gauge. Final conversion was completed in May 1892.

BOOKER, F. *The Great Western Railway: A New History*. David and Charles, Newton Abbot, 1977.

MACDERMOT, E. T. *History of the Great Western Railway*, vol. 1. Ian Allan, Shepperton, 1964.

14. Sonning Cutting

HEW 2182
SU 771 747 to
SU 747 740

In order to maintain the easy gradients he had set himself for the Great Western Railway main line to Swindon, Brunel had to penetrate Sonning Hill, which lay between Twyford and Reading. His original intention was to tunnel under Holme Park, but this was changed to a cutting 60 ft deep and 2 miles long on a line further to the south, for which a Parliamentary deviation order was obtained in 1837.

The contract for the works between Ruscombe and Reading was awarded to W. Ranger, but it became clear

J. B. POWELL

Sonning Cutting

early in 1838 that he would not complete in time. During the summer, when excavating in the clay should have been most straightforward, matters deteriorated further and the whole work was taken out of his hands in August and sublet in three parts. Knowles, the contractor who had taken the western end, also proved to be unsatisfactory and was dismissed in the October. The whole of the work was taken over directly by the Great Western Railway. R. Brotherhood, who had performed commendably in the centre portion, acted as principal subcontractor until completion of the earthworks. Between February and August 1839, 1200 men, 200 horses and two locomotives removed more than ½ million cu. yd of material. That from the western end was used to form embankments towards Reading, but that in the central part was tipped on the top of the cutting slopes. The predominantly Reading Beds which formed the bulk of the excavation proved to be prone to slips, and on Christmas Eve 1841 a slip caused a derailment resulting in the loss of eight lives. As late as 1954 a major rotational slip on the south side lifted the main up line and heavy remedial works were required.

The change from tunnel to cutting required the provision of two bridges, one of three brick arches carrying the

Great Western Turnpike Road (now the A4 trunk route), and one in timber carrying a secondary road. This timber bridge was to become a prototype form of construction and a familiar trademark of Brunel, particularly in Devon and Cornwall. The bridge (HEW 684) was reconstructed in 1891 with a wrought-iron centre arch and two side arches in brickwork. The north abutment of the A4 bridge settled in 1971 and remedial works were carried out. However, in view of the uncertainty concerning its long-term life a completely new bridge was built alongside, which now carries the eastbound traffic. It was the first post-tensioned, prestressed concrete, continuous box girder using an incremental launching system in Britain. Design was by Bullen and Partners, and construction by Reed and Mallik. The total costs were about £400 000.

BEST, K. H., KINGSTON, R. H. and WHATLEY, M. J. Incremental launching of Shepherd's House Bridge. *Min. Proc. Instn Civ. Engrs*, 1978, **64**, Pt. 1, 83–102.

MACDERMOT, E. T. *History of the Great Western Railway*, vol. 1. Ian Allan, Shepperton, 1964, 50–51.

15. River Thames Bridge, Great Western Railway, Windsor Branch

HEW 772
SU 961 773

Objections from Eton College and others had prevented the inclusion of a branch from Slough to Windsor in the Great Western Railway's Act of 1835. This opposition was overcome by 1848, when the Great Western Railway obtained parliamentary approval for the branch, subject to provision for the protection of the College amenities, including a requirement for the waterway to be kept clear, to the satisfaction of the College authorities, where the railway crossed the River Thames.

For this crossing, the Great Western's Engineer, I. K. Brunel, designed a wrought-iron bowstring arch bridge of 202 ft skew span. Mr. George Hennet was the contractor. The branch was opened on 8 October 1849. There are three main bowstring trusses, forming two bays for the rail tracks. They were originally carried on 6 ft diameter cast-iron piers filled with concrete, but these were replaced by brick abutments in 1908, which reduced the span to 184 ft 6 in. At the same time the cross girders and

J. B. POWELL

River Thames Bridge, Great Western Railway, Windsor Branch

rail bearers were renewed in steel. The approach viaducts, originally in timber, were replaced by brick arches in 1861–65. There is now only one line of rails on the bridge.

MACDERMOT, E. T. (revised by C. R. CLINKER). *History of the Great Western Railway*, vol. 1. Ian Allan, Shepperton, 1964, 97.

HUMBER, W. *Iron Bridge Construction*. Weale, London, 1864, 228–29.

16. Bourne End Viaduct

HEW 1295
SU 892 870

A single-track branch line from Maidenhead to High Wycombe was built by the Wycombe Railway Company and leased to the Great Western Railway. Opened in August 1854, it crossed the River Thames at Bourne End on a timber viaduct, which survived for 40 years before being replaced by the present steel structure.

There are three spans over the river, the main girders being double-leaf trusses 98 ft 3 in. long and 9 ft 6 in. deep. These are supported on brick piers at the river banks and on cylindrical steel columns in the river. The through-type deck comprises steel cross girders and rail beams, with a floor of wrought-iron plates. There are brick arch approach spans at each end of the viaduct.

MACDERMOT, E. T. *History of the Great Western Railway*, vol. 1. Ian Allan, Shepperton, 1964, 173.

Hook Norton Brewery

J. B. POWELL

17. Hook Norton Brewery

HEW 2178
SP 348 333

Established as a village brew-house in 1849 the brewery rapidly expanded during the remaining years of the nineteenth century, extensions being necessary in 1865 and again in 1872. External factors materially assisted its growth, particularly the construction of the Banbury and Cheltenham Direct Railway between the years 1874 and 1887 and the quarrying of high-grade haematite ore between Hook Norton and Banbury.

Such were the demands for its products that in 1898 a major six-storey extension was put in hand, which is the characteristic building to be seen today. The design was by William Bradford of London and the equipment was installed by Buxton and Thornley of Burton-on-Trent, who also built the now rare and much prized stationary steam engine. It is a single-cylinder 25 hp simple engine operating at 70–80 lbf/sq. in. fitted with Corliss valve gear and trip inlet valves. Steam is now supplied by an oil-fired boiler. In addition to pumping fresh water from the wells below the building, the engine pumps from the hop back situated on the ground floor to the flat cooler on the fourth floor and lifts malt also to the fourth storey prior to providing power for the malt crushing process. The transmission is a mixture of gearing and belt drives. Much of the interior loading is carried by cast-iron columns, some dating back to 1872, and the cold liquor tank is also in cast iron with sections bolted together. The brewery, which is still in private hands, has opened a well-appointed visitors' centre and to celebrate 150 years of brewing a full history has been published. Prior application must be made to visit the brewery.

ELDERSHAW, D. *A Country Brewery, Hook Norton 1849–1999*. The Hook Norton Brewery Company Limited, Hook Norton, 1999.

18. Bliss Tweed Mill

HEW 2331
SP 304 267

The Bliss family was associated with Chipping Norton for almost 150 years. From humble beginnings as cottage weavers, one of the family, William Bliss I, realised that the new machinery needed large and centralised premises with all the processes gathered into one complex. In the early years of the nineteenth century he moved into premises in the town which became known as the Upper Mill, and a second, in 1810, on the Common which became known as the Lower Mill. His son Robert took over from 1816 followed by Robert's younger brother, William, in 1839. The real expansion now took place. Although initially reluctant to take on these respons-ibilities, William nevertheless promoted a large-scale expansion of the works, taking advantage of the buoyant national economy. The Lower Mill was rebuilt in 1855 and new machinery was installed a few years later. At the same time William, along with other local businessmen, persuaded the Oxford, Worcester & Wolverhampton Railway to construct a railway from Kingham (then called Chipping Norton Junction Station) to the town, and the branch,

4½ miles long, was opened on 10 August 1855. The works now had a cheap source of coal, making steam power economically viable. Although a catastrophic fire burnt down most of the Lower Mill in 1872, William was able to maintain production by double shifts at the Upper Mill. Meanwhile, George Woodhouse, the architect from Bolton, designed a new mill, which is the present building. It embodied a number of fire-resistant features, such as cast-iron columns and brick jack-arches. However, the heyday was over and the Bliss family pulled out in 1896. With a succession of new owners the Lower Mill had mixed fortunes until production finally ceased in 1980. Fortunately, that is not the end of the story, as the gracious country-house styled exterior, with its characteristic chimney, is still there to be seen. In 1988 permission was given for change of use, and Edward Mayhew carried out a high-quality conversion into residential apartments, leaving *in situ* some of the original interior features and the exterior largely unchanged. The building may be viewed from a lay-by on the A44 and from footpaths that cross the adjacent meadow.

LEWIS, D. and WATKINS, A. *The Bliss Tweed Mills, Chipping Norton*. Chipping Norton Local History Society, 1989.

19. River Thames Navigation

HEW 2183
ST 980 995 to TQ 166 717

The River Thames has been a main artery for trade from Roman times, and while from the eleventh century its importance dictated regulation in matters of obstruction, little was enacted in terms of management. The needs of millers, navigation and riparian owners were frequently in conflict. Structures were confined to weirs, mills, fish traps and a few bridges and, apart from the effects of these, the river made its own course, shoals and shallows, adding to the difficulties of navigation. Boats negotiated weirs by means of flash locks, a deterrent to efficient transport and not unaccompanied by hazards. The considerable loss of water inherent in this system could, by itself, severely inhibit navigation upstream.

With the growth of riverside towns and the expansion of trade in the eighteenth century, various proposals were put forward for improvement, including canal bypasses. The Thames Commission, responsible for theriver from the source to Staines, opted for improvements to the river itself. Accordingly, in 1771 it promoted an Act for pound locks from Lechlade to Maidenhead, together with ancillary works of dredging and creation of towpaths. Initially, eight locks were constructed between Sonning and Maidenhead by 1773, followed by 14 locks to Lechlade by 1792. Incorporated within the latter were locks at Iffley and Sandford, built in 1632 under the aegis of the Oxford–Burcot Commission formed in 1624 following the passing of Act 21 of James I. They were bought out by the Thames Commission in 1789 for £600.

While these advances were useful for local traffic, progress in making the river easily or conveniently navigable throughout its length had not kept pace with canal developments, in particular the aim of the canal companies to have an efficient route to London. There were a number of proposals, but of the canals to the north and west only the Grand Junction achieved its through route to the Capital. In 1789 the Thames & Severn Canal had joined the river at Inglesham above Lechlade, and by 1796 the Oxford Canal also had a physical connection with the river at Oxford, which in turn had its connections with the Midlands and beyond. A little later the Wilts & Berks Canal joined the Thames at Abingdon, followed in 1816 by the Kennet & Avon Canal linking with the Kennet Navigation to Reading. Long-distance carriers who used the river found an unreliable waterway with no locks downstream from Maidenhead. Even as late as the 1830s a lot of traffic came up the Grand Junction and down the Oxford via Napton in order to avoid the Thames, but presumably the river current had some bearing on the choice.

There was some progress. A further group of locks was built between Cookham and Egham in the early part of the nineteenth century, together with six locks between Penton Hook and Teddington, constructed under powers obtained by the City of London. Nevertheless, the general state of the river and the financial position of the Commission had become so bad that it became the subject of a Royal Commission, this resulting in the Thames Navigation Act of 1866. The result was the formation of the Thames Conservancy to replace the

Commission and within the next few years there was a spate of lock and weir rebuilding.

In recent years changes have followed more quickly, and the management of the river is now under the umbrella of the Environmental Agency. Road and rail transport have largely taken over the commercial traffic, but as this has dwindled there has been a commensurate growth in private use. Using up-to-date technology the Agency and its recent predecessors have widened the range of services to meet the needs of the whole community that enjoys the river today.

SKEMPTON, A. W. Engineering on the Thames Navigation, 1770–1845. *Trans. Newcomen Soc.*, 1983–84, **55**, 153–76.

THACKER, F. S. *The Thames Highway*. Vol. 2: *Locks and Weirs*. David and Charles, Newton Abbot, 1968.

20. Virginia Water

HEW 2181
SU 9568, SU 9569, SU 9668, SU 9669, SU 9768, SU 9769

At the time of its construction in the early 1750s Virginia Water was the largest man-made lake in the country, covering 150 acres. It was the brainchild of the Ranger of the Great Park of the Duke of Cumberland and remained part of the private Royal Park until opened to the public in 1830 following the death of King George IV.

The lake was fed by water draining from the high ground to the north and Shrubs Hill to the south, and the brook forming the lake runs east towards the junction of the A329 (formerly the Windsor Forest Turnpike) and the A30 (formerly the Great Western Road). After leaving the lake the stream becomes the River Bourne, which enters the River Thames at Chertsey. Initially there was a small dam or pondhead at the western end of the present lake, but Cumberland had a vision of something much more extensive and placed a new pondhead near the eastern end of the park boundary. An artificial tumbledown, the cascade was incorporated at the sluice. The first reference to the lake occurs in a guide book of 1753. All went well until a storm on 1 September 1768 destroyed the pondhead and the lake.

No remedial work was attempted until August 1781, but new proposals included the Wick Branch at the north-east corner and to accommodate this the pondhead had to be moved further to the east. A private

J. B. POWELL

Virginia Water, The Cascade

Act of Parliament (Pitt's Estate Act 1782; 22 Geo. III c.9) covered these improvements, but negotiations proved long, tedious and expensive. Approval to carry out the work was given by the Treasury in April 1782. The Deputy Ranger, Thomas Sandby, designed the layouts for the pondhead, but while he was a competent architect and topographical artist it was unlikely that he had sufficient experience to design a dam 30 ft high and 200 ft long. A failure occurred in June 1785. There had been mitigating circumstances of heavy rains and periods of continuous frosts, but construction continued and eventually water flowed over the new cascade in March 1789.

The other engineering feature of note is the five-arch bridge leading into the Park from the A329 at Blacknest. Today's bridge is the third on the site, the first being of wood and built by Henry Flitcroft at the same time as the lake. Initially intended as a three-span bridge, a single arch of 165 ft was finally built, this being 20 ft wide with a clearance of 20 ft to water level. Although surviving the storm of 1768 it was derelict by the early 1780s and was replaced by a stone bridge designed by Thomas Sandby. Construction took place between 1783 and 1789. This bridge had a comparatively short life, being replaced by the present bridge designed by Sir Jeffry Wyatville at a cost of £9575.

The creation of the lake with its associated tree plantings has complemented the natural beauty of the area and testifies to Cumberland's grand design 250 years ago. The architects were T. Sandby and J. Wyatville (five-arch bridge).

ROBERTS, J. *Royal Landscape. The Gardens and Parks of Windsor*. Yale University Press, New Haven, CT, 1997, 34, 391–501.

SOUTH, R. *Royal Lake: The Story of Virginia Water*. Buckingham, 1983.

21. Oxford Canal (Banbury to Oxford)

HEW 1612
SP 362 845 to SP 508 064

The Act for the construction of a narrow canal from the Coventry Canal at Longford to Oxford was passed in April 1769. Work proceeded from the north end, and the canal was opened to Banbury in March 1778. Owing to shortage of capital work stopped there, but by an agreement made in 1782 at Coleshill between the Oxford and Fazeley and the Trent and Mersey companies the Oxford company was obliged to complete to Oxford. Accordingly, it obtained a new Act in April 1786 giving it greater borrowing powers and the work restarted. There were 28 locks, all with single top and bottom gates, and 38 of the characteristic lifting bridges. A slight change of route occurred at Heyford, where the canal used the bed of the River Cherwell and a new cut was made for the river alongside, thus saving two aqueducts.

The canal crossed the Cherwell at Aynho and again at Shipton, while to the north of the village the river and canal have a common bed for ¾ mile. One major change occurred north of Oxford, where the engineer, Samuel Simcock, put forward a proposal to take the canal through Kidlington rather than Begbroke, almost a mile to the west. At the Oxford terminus there were some difficulties in the negotiations, but the basin was finally opened on 1 January 1790. A warehouse and offices followed in 1796, the former being built by prisoners from Oxford Gaol at a cost of £3045.

Prisoners were also used to construct Isis Lock in the same year, which gave access to the River Thames.

Further to the north, near what is now the A44 Peartree Hill interchange, a 500 yd cut was made at the Duke of Marlborough's instigation to link the canal to the Thames, thereby avoiding King's Weir and Godstow Lock for westbound traffic.

In view of the poor state of the Thames and in an effort to reach London, the Oxford company supported the Hampton Gay scheme, named after a village north of Oxford, whereby an extension would be built, leaving the Oxford at Thrupp, across to Marsworth and then largely following the projected Grand Junction route through the Bulbourne and Gade valleys. Clearly Parliament could not countenance two canals on virtually the same route. The protagonists met but could not agree, and the Hampton Gay scheme came to nought. The engineers were James Brindley, until his death in 1771, and Simcock. James Barnes was Engineer for the final stage of construction.

SKEMPTON, A. W. *A Biographical Dictionary of Civil Engineers in Great Britain and Ireland*. Vol. 1: *1500–1830*. Thomas Telford, London, 2002.

COMPTON, H. J. *The Oxford Canal*. David and Charles, Newton Abbot, 1976.

22. Kennet Navigation and Kennet and Avon Canal

Canal:
HEW 1034
ST 754 644 to SU 470 672

Navigation:
SU 470 672 to SU 714 730

An Act for making the River Kennet navigable from Newbury to Reading was passed in 1715. John Hore was appointed Engineer and he constructed 21 locks together with several bypass cuts of the river. Opened in 1723, trade was sluggish until the Navigation was bought by Francis Page in 1767 who, with his sons, built up a thriving business.

The construction of the Kennet and Avon Canal between 1794 and 1816 linked with the Navigation at Newbury and this gave an impetus to trade. Clearly the Canal and Navigation were interdependent, so that their separate existence seemed to the canal company to be anomalous and the Navigation was incorporated in the Canal in 1812 for the sum of £100 000.

The opening of the Great Western Railway in 1841 rapidly affected the fortunes of the waterway and it was purchased by the railway in 1851. By the early part of the twentieth century it had largely fallen into disuse. However, through navigation from Bristol to London has been made possible again by an ambitious programme of restoration. The Engineer for the Navigation was J. Hore and the Engineer for the Canal was John Rennie.

CRAGG, R. *Civil Engineering Heritage: Wales and Western England.* Thomas Telford, London, 1997, 146–55.

HADFIELD, C. *British Canals.* David and Charles, London, 1984, 200–04.

Additional Sites

Numbers in **bold type** indicate Historical Engineering Works (HEW) site numbers.

1. Thames River Engineering

Battersea Bridge, **2342**, TQ 270 774

Battersea Rail Bridge, **2254**, TQ 266 765

Hampton Court Bridge, **2348**, TQ 154 685

Kew Bridge, **2345**, TQ 190 778

Lambeth Bridge, **2339**, TQ 304 799

Putney Bridge, **2344**, TQ 242 758

Putney Rail Bridge, **2255**, TQ 244 757

Richmond Bridge, **2346**, TQ 173 745

Vauxhall Bridge, **2340**, TQ 302 782

Walton Bridge, **2349**, TQ 093 666

Wandsworth Bridge, **2343**, TQ 259 756

2. Public Health

Addington Pumping Station, Croydon, **2223**, TQ 370 638

Arkley Water Tower, **2236**, TQ 222 957

Ashford Common Treatment Works, **2227**, TQ 087 697

Broadmead Pumping Station, **277**, TL 345 142

Hampton Pumping Station & Reservoirs, **2230**, TQ 127 693

Hanworth Road Treatment Works, **2228**, TQ 106 706

Island Barn Reservoir, **2232**, TQ 138 670

Mogden Sewage Treatment Works, **299**, TQ 155 750

Shooter's Hill Water Tower, **2235**, TQ 438 765

Surbiton Treatment Works, **2233**, TQ 173 672

Turnford Aqueduct, **277**, TL 362 047

Waddon Pumping Station, Croydon, **2224**, TQ 313 638

4. Rivers and Canals

Cast Iron Roving Bridge, Camden, **524**, TQ 287 841

Hanwell Lock Flight, Grand Union Canal, **1718**, TQ 149 796

Limehouse Basin, **1718**, TQ 364 810

Paddington Basin, **1718**, TQ 267 816

5. Roads and Road Transport

Coal Duty Posts, **1848**

Great West Road, **2260**, TQ 229 788 to TQ 217 781

Tollhouse, *The Spaniards Inn*, Hampstead, **2356**, TQ 266 873

Westway and the A40 (M) **2333**, TQ 274 817 to TQ 230 810

8. Notable Buildings

Brixton Windmill, **2300**, TQ 305 744

Canary Wharf Tower, **2337**, TQ 375 803

Lloyd's Insurance, **2364**, TQ 331 811

London City Airport, **2312**, TQ 425 805

Morden Snuff Mill, **2297**, TQ 263 684

Mumford's Mill, Deptford, **2296**, TQ 376 771

Penguin Pool, London Zoo, **2382**, TQ 283 834

Royal Horticultural Halls, Westminster, **2319**, TQ 295 789

Selfridges, Oxford Street, **2365**, TQ 283 811

Shirley Windmill, Croydon, **2302**, TQ 355 651

The Royal Mint, **2306**, TQ 358 803

9. The Thames Valley

River Ock Bridge, Abingdon, **363**, SU 496 966

Burford Bridge, **2367**, SP 252 125

GWR Transfer Shed, Didcot , **2372**, SU 522 908

Halfpenny Bridge, Lechlade, **628**, SU 213 993

Henley Bridge, **2371**, SU 963 827

Hertford Bridge, Oxford, **2369**, SP 517 064

Magdalen Bridge, Oxford, **2368**, SP 521 061

Munstead Water Tower, Godalming, **1454**, SU 988 428

Shillingford Bridge, **2370**, SU 597 920

Shiplake Viaduct, **1296**, SU 779 787

Swinford Bridge, Eynsham, **1915**, SP 443 087

Wallingford Bridge, **2366**, SU 610 895

London Memorials to Civil Engineers

London County Council/Greater London Council Blue Plaques

On houses lived in by the following:

BAZALGETTE, Sir J. W.: 17, Hamilton Terrace, Regents Park

BRUNEL, Sir M. I. and BRUNEL, Isambard Kingdom: 98 Cheney Walk, Chelsea

MANBY, Charles: 60 Westbourne Terrace

STEPHENSON, Robert: 35 Gloucester Square

Westminster Abbey

BAKER, Sir Benjamin (1840–1907): commemorated in a window on the north side of the nave.

BRUNEL, Isambard Kingdom (1806–59): commemorated with Trevithick in a window on the south side of the nave.

SMEATON, John (1724–92): floor plate placed in the nave.

STEPHENSON, George (1781–1848) and STEPHENSON, Robert (1803–59): commemorated in a window in the north aisle of the choir.

STEPHENSON, Robert (1803–59): buried in the centre of the nave (memorial plate), next to Telford.

TELFORD, Thomas (1757–1834): buried in the centre of the nave.

TREVITHICK, Richard (1771–1833): commemorated with Brunel in a window on the north side of the nave.

WATT, James: bronzed plaster bust in the Chapel of St. Paul.

WOLFE-BARRY, Sir John (1836–1918): commemorated in a window on the north side of the nave.

Other Sites

BAZALGETTE, Sir Joseph: bronze bust by George Simmonds, in a stone niche, Victoria Embankment facing *Northumberland Avenue*, unveiled in 1909 by the London County Council.

BRUNEL, Isambard Kingdom (1806–59): statue on a stone pedestal, Victoria Embankment, just east of Temple Station.

BRUNEL, Isambard Kingdom (1806–59): seated figure, adjacent to platform 1, Paddington Station.

CLARK, William Tierney (1783–1852): memorial plaque, north wall of the nave, St. Paul's Church, Hammersmith, with illustration of a suspension bridge.

STEPHENSON, Robert (1803–59): standing figure by Carlo Marochetti (1869), forecourt of Euston Station.

WALKER, James (1781–1862): bust by Michael Rizzello, on a pedestal at *Brunswick Quay*, Greenland Dock, unveiled by the President of the ICE in 1990.

Imperial College of Science, Technology and Medicine

A series of uniform commemorative plaques placed on the south wall of the Civil Engineering Building (Exhibition Road entrance) in 1987. The plaques are placed on the facade in chronological order of the date of the engineers' birth.

Early group of great engineers

SMEATON, John, FRS (1724–92)

JESSOP, William (1754–1814)

TELFORD, Thomas, FRS (1757–1834)

RENNIE, John, FRS

STEPHENSON, Robert, FRS (1803–59)

BRUNEL, Isambard Kingdom, FRS (1806–59)

Contemporary theorists

TREDGOLD, Thomas (1788–1829)

HODGKINSON, Eaton, FRS (1789–1861)

Practising engineers of the mid- and second half of the nineteenth century

CAUTLEY, Sir Proby, KCB, FRS (1802–71)

HAWKSLEY, Thomas, FRS (1807–93)

BAZALGETTE, Sir Joseph, CB (1819–91)

BINNIE, Sir Alexander (1839–1917)

BAKER, Sir Benjamin, KCB, KCMG, FRS (1840–1907)

Academic engineers of the period

RANKINE, William John Macquorn, FRS (1820–72)

UNWIN, William Cawthorne, FRS (1838–1933)

REYNOLDS, Osborne, FRS (1842–1912)

Engineers working in the twentieth century

GIBB, Sir Alexander, CB, FRS (1872–1958)

FREEMAN, Sir Ralph (1880–1950)

GLANVILLE, Sir William, CB, CBE, FRS (1900–76)

General Bibliograpy

Biddle, G. and Nock, D. S. *The Railway Heritage of Britain*. Michael Joseph, London, 1983.

Binnie, G. M. *Early Victorian Water Engineers*. Thomas Telford, London, 1981.

Bird, J. *The Major Seaports of the United Kingdom*. Hutchinson, London, 1963.

Boyes, J. and Russell, R. *The Canals of Eastern England*. David & Charles, Newton Abbot, 1977.

Brees, S. C. *Railway Practice: A Collection of Working Plans and Practical Details of Construction*. John Williams, London, 1838

British Bridges. Public Works, Roads and Transport Congress, London, 1933.

Brown, J. (ed.). *A Hundred Years of Civil Engineering at South Kensington*. Imperial College, London, 1985.

Chrimes, M. M. *Civil Engineering 1839–1889: A Photographic History*. Thomas Telford, London, 1991.

Clark, E. F. *George Parker Bidder*. KSL Publications, Bedford, 1983.

Clements, P. *Marc Isambard Brunel*. Longman, London, 1970.

Cossons, N. *The BP Book of Industrial Archaeology*. David & Charles, Newton Abbot, 1987.

Cox, R. C. and Gould, M. H. (eds.). *Civil Engineering Heritage: Ireland*. Thomas Telford, London, 1998.

Cragg, R. (ed.). *Civil Engineering Heritage: Wales & West Central England*. Thomas Telford, London, 1997.

Hadfield, C. and Skempton A. W. *William Jessop, Engineer*. David & Charles, Newton Abbot, 1979.

Halliday, S. *The Great Stink of London: Sir Joseph Bazalgette and the Cleansing of the Victorian Metropolis*. Sutton Publishing, Sutton, 1999.

Jackson, W. E. *Achievement: A Short History of the LCC*. Longman, London, 1965.

Jephson, H. *The Sanitary Evolution of London*. T. Fisher Unwin, London, 1907.

Kently, E., Hudson, A. and Peto, J. (eds.). *Isambard Kingdom Brunel: Recent Works*. The Design Museum, London, 2000.

Labrum, E. A. *Civil Engineering Heritage: Eastern & Central England*. Thomas Telford, London, 1994.

Norrie, C. M. *Bridging the Years: A Short History of British Civil Engineering*. Edward Arnold, London, 1956.

Otter, R. A. (ed.). *Civil Engineering Heritage: Southern England*. Thomas Telford, London, 1994.

Owen, D. *The Government of Victorian London 1855–1889*. Harvard University Press, Boston, MA, 1982.

Priestly, J. *Navigable Rivers, Canals, and Railways throughout Great Britain*. Longman, London, 1831 (David & Charles, Newton Abbot, 1969).

Rennison, R. W. (ed.). *Civil Engineering Heritage: Northern England*. Thomas Telford, London, 1996.

Richardson, A. E. *Robert Mylne, Architect and Engineer*. Batsford, London, 1955.

Rolt, L. T. C. *George & Robert Stephenson*. Longman, London, 1960.

Rolt, L. T. C. *Isambard Kingdom Brunel*. Pelican, London, 1970.

Rolt, L. T. C. *Thomas Telford*. Penguin, London, 1979.

Rudden, B. *The New River: A Legal History*. OUP, Oxford, 1985.

Schofield, R. B. *Benjamin Outram 1764–1805: An Engineering Biography*. Merton Priory Press, Cardiff, 2000.

Sheppard, F. *London 1808–1870: The Infernal Wen*. Secker & Warburg, London, 1971.

Simmons, J. and Biddle, G. (eds.). *The Oxford Companion to Railway History*. OUP, Oxford, 1997.

Sisley, R. *The London Water Supply*. Scientific Press, London, 1899.

Skempton, A. W. (ed.). *John Smeaton FRS*. Thomas Telford, London, 1981.

Smith, D. (ed.). *Perceptions of Great Engineers*. Science Museum, London, 1994.

Smith, D. (ed.). *Water-Supply and Public Health Engineering*. Ashgate, Aldershot, 1999.

Thacker, F. S. *The Thames Highway*. Vol. 1: *General History* (1914). Vol. 2: *Locks and Weirs* (1920). Reprinted by David & Charles, Newton Abbot, 1968.

Thomas, R. H. G. *London's First Railway: The London & Greenwich*. Batsford, London, 1972.

Vine, P. A. L. *London's Lost Route to the Sea*. David & Charles, Newton Abbot, 1986.

Watson, G. *The Civils: The Story of the Institution of Civil Engineers*. Thomas Telford, London, 1988.

Name Index

Engineers

Architects

Contractors

Others

Subject Index